I0026793

Management
and
Leadership

INSIGHT FOR
EFFECTIVE PRACTICE

Dag Forssell

Living Control Systems Publishing
Menlo Park, CA

Copyright © 1993-2013 by Dag C. Forssell

All rights reserved.

Numerous introductions, explanations and articles in PDF format, tutorials and simulation programs for Windows computer, links to other PCT web sites and related resources are available at the publisher's website www.livingcontrolsystems.com.

Library of Congress Control Number: 2008902333

Publishers Cataloging in Publication:
Forssell, Dag Carl, 1940–
     Management and leadership : Insight for Effective Practice
     /by Dag C. Forssell
     viii, 98 p. : ill. ; 28 cm.

     978-0-9740155-5-2 (softcover, perfect binding)
     978-1-938090-05-9 (hardcover, case binding)

     1. Control (Psychology). 2. Perceptual control theory.
     3. Management 4. Control theory.   I. Title.
     II. Title: Insight for Effective Practice.

     BF634-L4.   2008

v. 0513 —minor updates, recommended readings added.

⊗ The paper used in this book meets all ANSI standards for
     archival quality paper.

Contents

ABOUT THIS BOOK   v

ACKNOWLEDGMENTS   vii

INTRODUCTION   Why study perceptual control theory   1

PURPOSEFUL LEADERSHIP   Why a theory-based leadership program?   7

PERCEPTUAL CONTROL   A new management insight   13

PERCEPTUAL CONTROL   Management insight for problem solving   25

PERCEPTUAL CONTROL   Leading uncontrollable people   37

PERCEPTUAL CONTROL   Details and comments   49

SCIENCE   Are all sciences created equal?   59

COMPARISON   Chemistry versus psychology   75

BOOK JACKET   Behavior: The Control of Perception   79

SEMINAR INFORMATION   Purposeful Leadership   83

RECOMMENDED READING   PCT Literature & Resources—a flyer   93

This book

Management and leadership
Insight for Effective Practice

is available as a free PDF download from the publisher's website, www.livingcontrolsystems.com, as well as the free online libraries www.archive.org and www.z-lib.org, which will help ensure that this book and others on the subject of *Perceptual Control Theory, PCT,* will be available to students for many decades to come.

File name: ManagementLeadershipForssell2013.pdf

The file is password protected. Changes are not allowed. Printing at high resolution and content copying are allowed. Before you print, check the modest price from your favorite Internet bookstore.

For related books and papers, search *Perceptual Control Theory*

For drop ship volume orders, mix and match, contact the publisher.

For biographical information about Dag Forssell, see the publisher's website.

Minor updates and this note added in 2020.

About this book

This book introduces a new understanding of purposeful behavior and shows how to apply it to a wide range of leadership problems

Running a company, department or team has been more difficult than it needs to be because we have lacked an understanding of human behavior that actually fits the way human beings work.

Perceptual Control Theory (PCT) gives an intuitively satisfying explanation for purposeful human behavior, also known as control. Hierarchical PCT (HPCT) outlines a hierarchical arrangement of multiple control systems as a testable explanation that allows for the complexity of our experience.

PCT focuses on how we look at and experience things, and the way these perceptions are compared with experiences we want. PCT explains how thoughts become actions and feelings and why stimuli appear to cause responses. PCT improves our understanding of human interpersonal behavior, including conflict, cooperation and leadership in families, education, business and society.

Conflict is the root cause of most management problems. Not only does it waste energy—it destroys cooperation, teamwork, personal initiative, care, productivity and quality. Failure to resolve conflict results in stress, frustration, resentment, destruction of personal relationships, and personnel turnover.

In this book you learn that we *are* controllers, that it is our nature to control, and that when we attempt to control others we easily create conflict.

PCT shows what control is and how it works; how it gives rise to conflict or cooperation, depending on what individuals want and how they see things. Control is not a dirty word. Control is necessary for life and being "in control" or contented is satisfying. When others attempt to control us we resist and dislike it.

Control, conflict, and cooperation are illustrated. Consistent application of *mapping and influencing wants and perceptions* to conflict resolution, team development and non-manipulative selling makes uniform leadership practice possible—leading to high performance, consistent results, and mutual satisfaction. Insight from HPCT is applied to vision and mission statements and TQM. You can see how PCT and HPCT turn a "soft" subject into a "hard" science.

We are all psychologists; we all deal with other people. Our understanding and skill determines our effectiveness and satisfaction as leaders, managers, salesmen, teachers and friends, both in the workplace and in our personal lives.

With PCT, leaders and staff can learn the same testable understanding and effective approach. You deal with your associates at all levels just like they deal with customers and suppliers. When you understand PCT, dealing with people no longer will be complex and confusing, a matter of luck, a gift, or something best left to specialists.

Acknowledgements

Several friends have supported me as I developed a training program and papers to explain PCT and suggest applications.

The seminal insight and writings of William T. (Bill) Powers are the solid foundation on which my work is based. Bill Powers is a warm human being who walks his talk, an untiring champion of clear thinking and a patient teacher. Members of the Control Systems Group, an association of researchers exploring PCT, have helped me keep every phrase as correct, clear and unambiguous as possible in order to avoid misinterpretation and confusion with conventional thinking.

Translating the elegant framework of human understanding we call PCT into bite size pieces of explanation and direction for everyday life has proven a challenge. Ed Ford has traveled this path before me and has written about how to improve personal relationships. Applying insight from the first principles of PCT, Jim Soldani was able to effect lasting improvements in the performance of a manufacturing plant.

Based on traditional research and personal experience, Mike Bosworth has developed and teaches a non-manipulative sales program called Solution Selling® which fits well with conclusions drawn from PCT. Mike's suggestion that I develop a program to teach sales managers how to develop and maintain productive personal relationships with salesmen got me started on my mission to explain and illustrate PCT.

The original versions of the first three papers in this book were published in the Engineering Management Journal. My editor, Dr. Ted Eschenbach, made many helpful suggestions for clarification, especially where PCT leads to conclusions that surprised him.

Dag Forssell March, 1995

Engineering Management Journal (EMJ)

Engineering Management Journal is the quarterly journal of the American Society for Engineering Management (ASEM). It is designed to provide practical, pertinent information that relates to the management of technology, technical professionals, and technical organizations. . . .

Information about EMJ: www.asem.org

WHY STUDY PERCEPTUAL CONTROL THEORY?

Why study perceptual control theory?

What's in it for you?

Are you curious why and how people do what they do? Would you like to be more effective as a parent, teacher, manager, spouse or friend—and develop more satisfying relationships in the bargain? I bet you will discover that you will gain more useful, dependable insight more quickly when you learn Perceptual Control Theory (PCT) than you possibly could any other way. You will begin to question many conclusions that you previously thought were well-established truths.

I am a mechanical engineer who came to the United States from Sweden in 1967 with my wife Christine. My curiosity about "what makes people tick" was aroused when Christine became a salesperson in 1976. I began to study sales, management, public speaking, listening skills, parenting and psychology. I thought a book or program was worthwhile as long as I found an idea or two that made sense to me and that I thought I could use.

In 1988 I came across *Behavior: The Control of Perception*[1] by William T. Powers. I soon realized that this book outlined a new scientific approach to understanding human nature—it was not just another pop-psychology or self-help book with one or two good ideas.

As I studied PCT, I saw an entirely new way to explain what behavior is and what actions accomplish. PCT looks at behavior from the inside perspective of the behaving person, not from the outside perspective of an observer. PCT shows clearly that actions are rarely deliberate; a person is not necessarily aware of actions. Actions influence the environment (or attempt to) so that a person experiences what the person wants to experience at the time and under the circumstances.

1) See page 81 for a reproduction of the book jacket.

With PCT insight, I now see actions as symptoms of wants and understandings and ask people about their wants whenever a conflict arises. In PCT-speak, this means that I ask them what the situation looks like from their inside perspective and what perceptions they are trying to control, rather than jump to conclusions about the situation based on my incomplete observations from the outside, supplemented by a generous helping of other information retrieved in real time from my personal store of understanding and memories—in other words, based on what I imagine.

I realized that I had on many occasions caused conflict with others by insisting on my interpretations and by trying to impose my wants, telling people what to do and how to do it. So now I do my best to offer information instead, information that my friends and associates can consider and make their own; information that will affect how they understand their world, change what they want—and thus change their actions.

As Christine and I began to apply this understanding in our own interactions, our already good marriage became even closer. If one of us is upset about something, we let the other know we have a strong error signal. This leads our conversation directly to a discussion of a want (the reference signal), compared to a perception or interpretation of what is (the corresponding perceptual signal). This approach eliminates the oh-so-intuitive focus on actions. It removes any accusatory tone from discussion and helps us support each other by reviewing the want—it's origins in higher-level understanding, appropriateness and selection, stored perceptions (imagination) mixing with current input, creating our current perception or interpretation of what is, actions we have tried, and unintended consequences of each other's actions. It becomes easier to make suggestions and accommodate each other's preferences. We recognize that persistent error signals cause reorganization and can be harmful, but accept the idea that error signals and reorganization are part of life.

I now put my understanding to use daily when dealing with customers—anticipating what perceptions they are controlling—and find myself getting along much better than I did earlier in my career.

My whole outlook on life has changed and I feel much more accepting and at peace with myself than I used to, all because I have gained a fundamentally different understanding.

The remarkably simple explanation developed by Bill Powers is based on both the principles and methods of successful physical science and it remains consistent with our intuition about the autonomy and complexity of human nature. Once you understand this explanation, you will find it both elegant and compelling. The explanatory mechanism introduced by PCT is testable through various experiments, so don't accept it on anyone's authority. Test it for yourself—every step of the way. You will find that PCT covers much ground and explains a great deal of our experience, but leaves many mysteries for future researchers to explore, such as consciousness, awareness, attention and memory—mysteries for which no-one has any definitive answers.

When you study PCT, bear in mind that this is not just an idea of the month, another passing fad or "The Powers Philosophy," but a simple, basic description of the marvelous mechanism that *is* a human being, always has been, and always will be. You *are* a perceptual control system, as is every living being. That is why it is important to understand how a perceptual control system works, and this is why we offer tutorials and simulations you can run on your own computer.

When you understand the mechanism described by Perceptual Control Theory and see that people always control perceptions, you can understand any new interaction by reasoning based on PCT. You no longer need to memorize advice for all possible circumstances. Social interactions in all their apparent complexity suddenly become much simpler and easier to understand. This kind of insight you cannot ever learn from descriptive science—a storytelling or "this is what you do" approach to learning.

Understanding the basic mechanism will only be the beginning of your personal transformation. As you live through new experiences, you will naturally examine them in the light of PCT. Over time, your understanding will mature and flavor your entire outlook on life.

Why worry about explanations?

PCT offers an explanation. Why should you care about an explanation? I have heard many people say: "Don't confuse me with theory, tell me what to do!" I think that there is good reason for this doubting attitude when it comes to education that deals with social interaction. Explanations come in many flavors. Some are vacuous, some superfluous, some erroneous and some very useful indeed, providing solid understanding and structure for the way we think. Let me briefly[2] share some thoughts on explanations and science:

Explanations are not necessary to live

Fishes, cats and people get along just fine without any explanations at all. We all learn from experience. We want something and act in various ways until we experience what we want. Then we remember what we did (or rather, what perceptions we were controlling at the time).

---

2) See *Are all Sciences Created Equal*, pages 59–74.

Some explanations amount to conversation

Explanations sometimes merely restate the problem (you can't read because you are dyslexic, where dyslexic is Greek for "can't read"), offer conversational speculation (the customer bought from you because he liked you best), or lump symptoms together in groups to define a "syndrome" which provides an illusion of scientific understanding.

Learning from experience provides little structure

Learning from experience, you deal with each situation as it occurs. As you accumulate experience, you say: "In these circumstances, do that." It takes a very long time to accumulate a variety of experiences and attempt to draw general conclusions from them. Unless you happen to hit on some very solid generalizations you will likely be surprised over and over when things don't turn out the way you expected. Your generalizations are unlikely to provide dependable structure for your thinking and guidance for new and different situations.

Many widely accepted explanations are wrong

Our language is full of references to the idea that the environment and people in it make us do and feel things. "You make me so angry!" "Look what you made me do!" "Our managers reinforce desirable behavior." "I want to make you happy." "His reaction is understandable when you know how he has been conditioned." We have all grown up with these concepts and explanations and they sure can seem valid when you look at people's actions from the outside. Nevertheless, the Stimulus-Response concept of linear causation is simply wrong, and the concept of the brain issuing detailed commands, likewise linear causation, is also wrong. Neither is physically feasible. Statistical findings, resulting from research based on these intuitively appealing concepts, are most often of very low quality.

Languages are made up of explanations

The language of a particular science at any point in time defines concepts, explanations and functional relationships in a coherent whole. The language and its concepts determines how we view and describe what we experience. When you have learned a scientific language it becomes very difficult to step outside it and see an entirely different explanation, based on different basic concepts, where words take on different meaning. What you already "know" seems "right" and different explanations seem "wrong."

In his book *Inventing Reality: Physics as language* (NY: Wiley, 1990), Bruce Gregory reviews successive languages in the physical sciences, each one replacing its predecessor. When a new, more useful, testable and demonstrably more valid language is radically different, a scientific revolution has to take place eventually, because the old explanations and concepts lose their validity when compared to the new.

Scientific revolutions happen

I changed my notions about scientific progress when I read *The Structure of Scientific Revolutions* by Thomas S. Kuhn (Univ. of Chicago Press, 1970). I had thought that scientific progress always meant adding new discoveries to an already validated body of knowledge. Now I understand that the history of science is a history with long spells (many decades or centuries) of knowledge accumulation, punctuated by intellectually violent transitions where old knowledge is superseded by new concepts that give rise to new detailed explanations. Sciences start over. I am happy to particiate in a movement that is bringing a fundamentally new, testable and very practical explanation to the life sciences.

Good explanations make a huge difference

In-depth explanations provide structure

With a structure of in-depth explanations, such as provided by the contemporary engineering sciences, you can extrapolate from known principles and designs to completely new, never before attempted, actions and designs—yet be very confident things will work out. Such a body of in-depth explanations become a way of thinking—a systems concept in PCT language. This structures your thinking and provides a framework by which you fit additional experiences and conclusions into a coherent understanding. PCT offers a structure by which you can organize your understanding of living organisms and make sense of their behavior.

Where explanations prove correct – science can progress

The impact of correct, useful explanations is readily seen in the recent history of the physical sciences. New concepts, a new approach to measurement and a new set of physical explanations were introduced by Copernicus, Galileo, Kepler and Newton in the 1500s to 1700s, laying the foundations for modern physical science and the remarkable progress we have benefitted from during the last 300 to 400 years.

When students learn about the physical sciences today, they replicate many fundamental experiments and accept the theoretical explanations that go with them because they can see near perfect agreement between their own experience and the explanation. When engineers design devices today, they confidently expect them to work as predicted.

PCT offers a correct explanation – science can progress

When you learn about PCT today, you can replicate many fundamental experiments, run the simulations and accept the explanation that goes with them based on your own judgement, because you can see near perfect agreement between your own experience and the explanation. When you offer your friends information passed through the filter of PCT understanding, you will be offering better (and less confusing) information than they can get with today's descriptive languages and they will be able to control their perceptions better than they do now—they can be more satisfied. When you deal with people in the future, you will have greater understanding and confidence. You will be able to bring some order out of apparent chaos in your personal world.

Dag Forssell July, 1997, revised 2003

Purposeful Leadership — Why a Theory-Based Leadership Program?

Purposeful Leadership — why a theory-based leadership program?

The value of a good theory

Kurt Lewin, researcher of group dynamics at MIT, said:

> There is nothing as practical as a good theory.

In the engineering and physical sciences, this is well established. Engineers and physical scientists recognize that a good theory allows for the prediction of performance long before actual experiment or production. Good theories have allowed us to communicate, understand and produce better than ever before in history. Good theory bolsters our common sense by providing a clear framework for understanding.

When it comes to the important area of human affairs, the situation is very different. Many theories have been offered over the years, attempting to explain human action, but none have measured up to scientific scrutiny the way theories do in the physical sciences. This is why many psychologists say that their theories and practices have nothing to do with each other.

Existing training programs

Companies spend millions of dollars on training relating to human affairs. To illustrate the variety, this list is taken from a recent CareerTrack® brochure:

Team Building: How to Motivate and Manage People™
Getting Things Done™
Making Meetings Work™
The One Minute Manager Live!™
Selling Smart™
How to Delegate Work and Ensure it's Done Right™
Assertiveness Training™
Personal Power™
How To Give Exceptional Customer Service™
Negotiate Like The Pros™
How To Set And Achieve Goals™

Stress Management For Professionals™
Controlling Anger™
How To Deal With Difficult People™
Self-Esteem And Peak Performance™

Programs like these may each be very good, but none are based on a validated theory of human behavior. They are based on "what seems to work"—the author's personal experience supplemented with anecdotal experiences and interpretations drawn from many sources. In the absence of proven theories in the area of human affairs, these programs cannot offer a universal framework of explanation. The focus and quality varies. Without a common explanatory framework, programs may be contradictory, even within themselves.

These multiple programs, offering multiple scenarios and entertaining stories, suggesting multiple prescriptions for "what and how to do," make understanding and dealing with people far more complex than it needs to be.

In most cases, people have fun and like the training, but four or five months later, little has changed in the workplace. I believe a reason for this is that most training is situational or anecdotal and focuses on "what and how to do."

Each participant is left to integrate the many disparate lessons of the training experience into the framework of their personal *understanding*, such as it is.

People want "practical" seminars focusing on "techniques," "skills" and "tools." This is all they have ever been offered, because absent a good theory = functional explanation, that is all anyone can deliver.

People ask: Show me what to do (cause) so I will get results (effect). This is fallacious, but that does not change the fact that this is what many people have come to expect and want.

Specific instructions on "what and how to do" are valid only in a given set of circumstances. Typically a training scenario is carefully selected and told with drama and humor by a speaker. You are told what the circumstances were, what the prospect was thinking, what was done and what the results were. You imagine that the same thing will happen if you do the same thing. You feel euphoric as you imagine success.

A large part of the "what and how to do" training does not really apply in individual cases because the world is full of varying conditions and changing disturbances. Lessons become irrelevant and are soon forgotten. Euphoria fades.

If much training is ineffective, how can Purposeful Leadership have lasting value?

Theory-based education and training

The major strength of the Purposeful Leadership program is that it explains and applies a new theory called Perceptual Control Theory (PCT). PCT recognizes and explains the phenomenon of control. PCT explains why and how people do what they do. PCT is based on neurology and clear, detailed and tested engineering concepts. PCT requires and offers scientific rigor with explanation and prediction.

PCT is a "hard" engineering science of psychology that is easy to understand for anyone who pays attention to the detailed, functional tutorials and simulations.

Once the phenomenon of control is observed and the detailed explanation understood, it will be seen that control is the fundamental organizing principle of life. Control is pervasive and can be seen operating at microscopic levels as well as at the macro level of human activity.

PCT explains a wide variety of phenomena of everyday experience because it goes beyond the predominant focus on cause and effect to explain the consistency of outcomes and the variability of means.

Specific instructions on "what and how to do" are valid only in a given set of circumstances.

Speed, Cost, Effectiveness

Instead of using **multiple** programs, each one covering some aspect of human interaction, you can use **one** to understand yourself and others in some detail. Participants can decide that the theory is good by testing it in their own lives. Everyone can draw conclusions from the same theory, supplemented and adapted with specific information as required by special applications. The time and expense of training is dramatically reduced. This **one** education is effective even as jobs are rotated, because a good theory applies everywhere (if it is really good and valid).

If this new theory is so much better, why is it not widely known already?

One reason is that it is new. Another is that this theory is dramatically different from the prevailing "soft" descriptive science of psychology. It causes a scientific revolution.

Scientific revolutions

The late Thomas S. Kuhn, leading scientific philosopher, professor at MIT and author of *The Structure of Scientific Revolutions,* (1970, University of Chicago Press) explained. From the book cover:

> ....Thomas S. Kuhn wastes little time on demolishing the logical empiricist view of science as an objective progression toward the truth. Instead, he erects from the ground up a structure in which science is seen to be heavily influenced by nonrational procedures... ....Science is not the steady, cumulative acquisition of knowledge that is portrayed in our textbooks. Rather it is a series of peaceful interludes punctuated by intellectually violent revolutions ....in each of which one conceptual world view is replaced by another....
>
> *Nicholas Wade, Science*

In this book, Thomas Kuhn introduced the term *paradigm* and suggested that scientists schooled in a certain set of views adopt them as their personal paradigms, then view the world through these paradigms—as if they were eyeglasses filtering information. The word paradigm means pattern. It is used to signify how we interpret a phenomenon; how we explain the world to ourselves.

In Kuhn's view, everyone is a scientist, and every world view might be called a personal science. Everyone has some framework of ideas of how the world "works" and views the world through those personal paradigms.

When a radically new theory is presented, understanding can be difficult. If the old and the new concepts are incompatible, it becomes very hard to see the new paradigm/science through the eyes of the old paradigm/science.

A reading of Kuhn's book makes it clear that there is lots of room in the sciences of today for coming revolutions.

New information—on any subject—is always interpreted using what you already understand. PCT itself explains why this is so. Where a person has existing convictions, conflicting information is either not comprehended or rejected. A person without convictions on a certain subject is more open to new information.

This is why scientific revolutions typically originate from outside the scientific community which has accepted the present paradigm.

Perceptual Control Theory

Perceptual Control Theory is a new "hard" engineering science of psychology. It offers description, causal explanation and prediction. Explanations which yield predictions with 99⁺% experimental confirmation are possible and are expected in time. Much development work remains to be done. Tests to date show 95-98⁺% correlation in simple experiments (which anyone can duplicate), with the remaining 2-5% accounted for by expected imperfection of control: less than infinite loop gain, slow response and loose or weak feedback through the environment.

PCT makes possible a transition from a "soft" empirical and descriptive science of psychology where theory and application are worlds apart, to a "hard" engineering science of life and psychology, where theory and application fit like hand in glove.

PCT requires a major shift in psychological thinking from the traditional approach. The traditional view considers behavior a dependent variable. PCT goes beyond this view. What is controlled is not behavior, but *perception*—what a person experiences. This is hard to grasp and accept for persons schooled in the version of the scientific method used in psychology and has slowed but not prevented publication. PCT itself explains why.

On the other hand, PCT is immediately acceptable—intuitively obvious—to people without traditional training, understanding and convictions. It is easy to understand and immediately useful.

Without a good theory, every problem must be solved by trial and error. You have to learn a lot of rules for every conceivable circumstance.

With a good theory you learn the theory, (as in physical science), then work problem after problem to learn to recognize how the theory applies and get used to using it. What you remember is the theory, not individual solutions.

With Purposeful Leadership and PCT, you learn a good theory, then spend time with application after application to recognize how the theory applies and get used to think that way. What you remember is the theory, not individual "what and how to do" solutions.

The power of a program based on a good theory is awesome. There is nothing as practical as a good theory-based program.

Dag Forssell November, 1992

What psychological insight do *you* think a management program should be based on?

Here is my take on the difference between the linear cause-effect theories behind most management programs and the theory behind Purposeful Leadership:

Feature	Contemporary life science	Perceptual Control Theory (PCT)
Description	*Yes*	*Yes*
Descriptive non-explanation	*Yes*	*No*
Successful functional explanation	*No*	*Yes*
Prediction = repeat of observation	*Yes*	*No*
Prediction = logical result of principles	*No*	*Yes*
Resorting to statistics, suggesting illusory functional relationships	*Yes*	*No*
Reasoning from first principles	*No*	*Yes*

PERCEPTUAL CONTROL —
A NEW MANAGEMENT INSIGHT

Number one in a series of three articles on PCT.
An early version of this article appeared in
Engineering Management Journal Vol. 5 No.4 Dec 1993

Perceptual control — a new management insight

ABSTRACT

This article discusses how leadership of an organization depends on psychology and introduces a scientific approach to psychology called Perceptual control Theory (PCT).

Using PCT, managers can learn how to encourage associates to align their wants with the goals of their organization. This gives managers the capability to develop cooperation and resolve conflict.

INTRODUCTION

Individual commitment to the job and interaction with others are both critical at all levels and in all areas of an organization. This is why management philosophy, management skills, and motivation are important topics. But one management approach has replaced another with such regularity that each new one is now greeted as a fad, to be suffered by the rank and file until it too is found wanting and yet another is tried in its place.

I suggest in this article that running a company, department or team has been more difficult than it needs to be because we have lacked a theory of human behavior that actually fits the way human beings work. Managers have a working acquaintance with human nature, but it is mostly based on accumulated practical experience, as journeymen of old based their knowledge of machines and materials on the accumulated lore of their trades. Modern engineering is what it is today because of theories: theories of matter and energy that permit accurate prediction of the behavior of material things. Until recently, there were no equivalent theories of human behavior.

This article introduces a new set of principles: the principles of control. Traditionally, control as an engineering discipline has been studied from the outside perspective of an observer who can see what is happening and make adjustments to the control circuit as needed. To show how the principles of control work in humans, we adopt a new perspective: from inside the controlling system, where all we know is what we perceive. This changes the flavor of the word *control*.

Defining perceptual control

People dislike the word *control*. To many, it is synonymous with manipulation and coercion. This is due to an incomplete understanding of how control works. Complete understanding shows how to eliminate manipulation.

Control is a pervasive natural phenomenon that has not been clearly described until well into this century. When you understand control, you can observe control processes in the behavior of bacteria, plants and animals all around you, all the time.

To control your perception means to influence (act on) your environment so that you perceive the environment the way you want to perceive it, according to some specification, want, or goal you set. To survive, an organism must successfully control several aspects of its environment.

You eat to influence your blood sugar level until your perception of it agrees with your specification for it. You pull the covers tighter on a cold night to influence the temperature of your skin so that it agrees with the warmth you want to feel. You work to influence your environment in many ways until your perceptions are to your liking. When they are, you call yourself content or satisfied. When you cannot influence your environment effectively, you experience stress. This may be due to an inability to act, because you want incompatible things at the same time, or due to conflict with someone else who influences the same aspect of your common environment, but with a different goal.

All behaviors are part of perceptual control processes. Our actions influence some aspect of the world as we perceive it. The word *control* is appropriate to describe the process. Control is necessary for life. Good control of one's own environment is satisfying. Control of one's environment by way of physical coercion and threats to others may be satisfying to oneself but is certainly not to the others; thus the general dislike of control. An understanding of control shows us how to control well to satisfy our own wants without violating the rights of others to control and satisfy their wants. An understanding of control enables us to develop productive and satisfying cooperation.

Results

To demonstrate the power of this new theory, let us briefly review results in a manufacturing group of 120 people made up of 60 assemblers and 60 support staff, part of a Fortune 500 company (Soldani, 1989). The performance of this group was ordinary, with various delays early in the month and overtime late in the month to make up for them.

The Operations Manager applied the new PCT insight with a questioning technique that encouraged everyone to align their wants with the single goal of completing work orders on schedule. In seven months, the measure of this goal went up from 23% to 98%, and other measures improved too. Overtime declined from 12% to 3%. Quality went up by a factor of 5. Work in process inventory fell by a third. Productivity went up 21%. Customer satisfaction, sales and morale—all went up. The group won "site of the month award" 11 out of the 12 months after that. Total savings added up to about 1.5 million dollars a year.

Organizations as control systems

Quality pioneer Dr. W. Edwards Deming talked of an organization as a system that must have an aim, where aim is the goal or specification of an intended outcome. This clearly implies that an organization is a control system. An organization can be portrayed as an interconnected hierarchy of control systems, in which each level of executives translates broad goals from above into more detailed functional goals for their own and lower levels. Ultimately, goals are translated into action by individuals who deal with the tangible outside world. (Exhibit 1). Reports about the outside world travel back up the chain of command, but are not shown in this illustration.

Drawing the organization chart this way, I distinguish between the inside and outside of the company's "brain." Inside is the thinking part. Outside is the physical world and actions in it.

I draw the organization chart this way for later comparison with the diagram of a person as a control system. In the organization, people at all levels act in three areas:

1. In the hierarchy of the "brain" portrayed here
2. At the interface, requesting action on the "outside"
3. On some outside quantity or process.

Ideally, people and teams respond to changing circumstances with their own initiative and ingenuity to achieve and maintain their assigned (and changing) goals; they stay in control. As a manager, all you have to do is give people the information and resources they need to do their jobs. If that does not work, the traditional view is that there is something wrong with them, not with the manager.

This may be a useful portrayal, but it is misleading to act as if an organization or any other social system is a true "control system," without carefully understanding the components. The implication of this metaphor is that the organization and the people in it automatically respond with their best efforts to simple commands issued from higher-level managers. This rarely happens. People work more side by side than in a rigid hierarchy. Dr. Deming was wise to also suggest that the aim of the organization must be clear to everyone in the organization.

People in a social organization are not dedicated components, as parts in a machine are. Everyone strives to satisfy a multitude of individual purposes, which may or may not include the goals requested from above. People are not on the job or attentive all the time. People do not react with predictable outputs to given inputs. They perceive differently and misunderstand each other. Capabilities vary. People develop internal and external conflicts that interfere with their commitments to the organization.

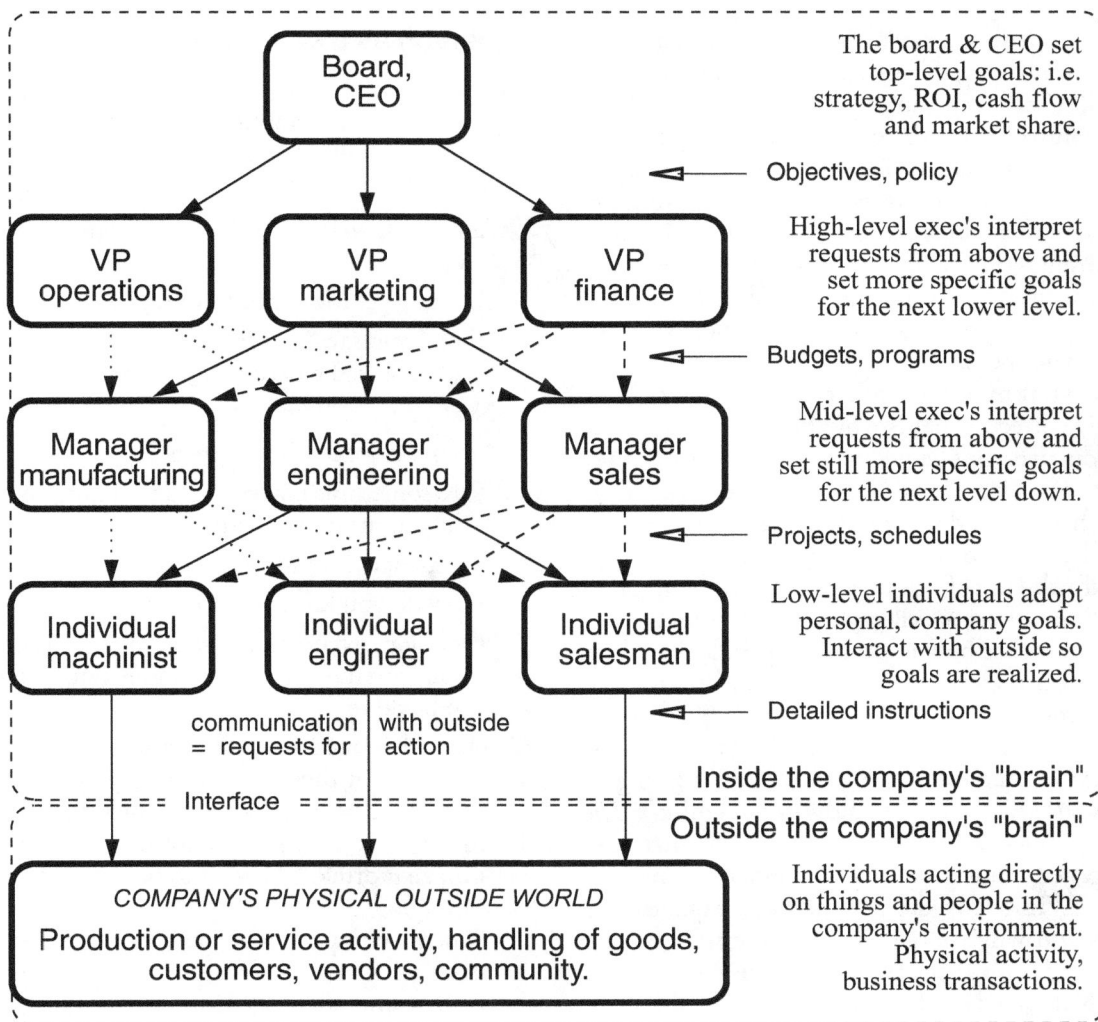

The board & CEO set top-level goals: i.e. strategy, ROI, cash flow and market share.

◁—— Objectives, policy

High-level exec's interpret requests from above and set more specific goals for the next lower level.

◁—— Budgets, programs

Mid-level exec's interpret requests from above and set still more specific goals for the next level down.

◁—— Projects, schedules

Low-level individuals adopt personal, company goals. Interact with outside so goals are realized.

◁—— Detailed instructions

Inside the company's "brain"

Outside the company's "brain"

Individuals acting directly on things and people in the company's environment. Physical activity, business transactions.

Exhibit 1. An organization as a hierarchical control system.

(Reports not shown).

Leadership depends on psychology

Many rules about management flow naturally from the basic assumption of conventional psychology—which we have all learned in school—that what people do depends on what happens to them. This assumption is called the linear "cause-effect" model of behavior. In psychoanalytic psychology, the cause-effect model shows up as the idea that past traumas are the cause of present emotional problems. In behavioristic psychology, the cause-effect model shows up as the idea that environmental stimuli (and reinforcements) are the cause of complex human behavior. In cognitive psychology, the cause-effect model shows up as the idea that internal stimuli (plans) are the cause of action—sometimes called controlled output or controlled behavior.

The linear cause-effect model and experimental method form the unspoken basis of all contemporary scientific psychology (Exhibit 2).

The linear "cause-effect" model itself, when applied to people, is incomplete and misleading because it describes only one aspect at a time of the relationships between people and their environment. It is true that what people do depends on what happens to them or what they plan, but it is also true that what happens to people depends on what they do.

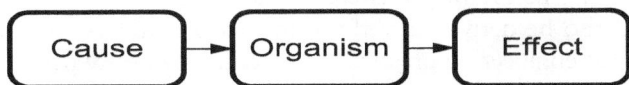

Exhibit 2. The cause-effect model.

Behavior is part of a circle of simultaneous and continuous cause and effect that has no beginning and no end. What occurs in this circle is a phenomenon called control—the process of producing predictable results in an unpredictable and changing environment. Control can be as simple as keeping your car in its lane on a windy day or as complex as keeping your business profitable in a shifting economy. Control is the process of carrying out and maintaining a purpose, such as driving or making money. Control is purposeful behavior.

Behaviorism and cognitive psychology are both partially correct. That is why they are so suggestive and seductive. But being only partially correct, they cannot offer a functional in-depth explanation. That is why neither works as a theoretical foundation for psychological or management practice.

In the absence of a complete theory, contemporary psychology offers a collection of observations and prescriptions for how to deal with people, based on trial and error by many practitioners, collected over a long time.

Even the best contemporary leadership programs, such as Dr. Stephen R. Covey's *7 Habits of Highly Effective People* or Dr. W. Edwards Deming's *The Deming Management Philosophy* can only offer a set of suggested "principles" (rules of conduct or prescriptions). Without in-depth functional explanations of how and why people behave, these rules will be interpreted with extra large variations by different individuals based on their own personal experiences. Consistent management understanding is hard to attain, and any such management program is hard to implement in an organization.

This is not a criticism of leadership programs, but a comment on the lack of understanding in our society. The rules of a wise leadership program are by definition supposed to be aligned with the underlying functional relationships that govern the behavior of people. But without knowledge of those functional relationships, how can you tell the wisdom of one rule from another, contradictory one? How can you select a rule when it is unclear how it fits the present situation? We would all be better off if the functional relationships *themselves* were laid bare. Such knowledge can only strengthen a wise leadership program. An effective leadership program can also be derived directly from such knowledge, just as engineering practice can be derived directly from proven engineering theories.

A person as a control system

Introducing Perceptual control Theory: Contrasting with the basic assumption of conventional psychology stands the intuitive recognition wise managers often develop—that *people are control systems.*

> Ask any manager: "What does control mean to you?"
>
> You may get the answer: "Oh, power."
>
> "Fine," you say, "you can't control anything if you don't have the necessary power."
>
> Next you ask: "What is your power for?"
>
> "To make something happen."
>
> "OK, something is what you are working on. What do you mean 'happen?'"
>
> "To change things so I meet my goal."
>
> "How do you know what to do?"
>
> "Oh, I look where things stand now and compare that with the goal. Then I figure out what needs to be done."

You have described a simple control system. People understand that they control. Many of our sayings reflect this: "If you really want something, you will find a way." "You can lead a horse to water, but you can't make it drink." "I want to be free. . . to do my own thing."

When we respect other people, we respect their understandings and beliefs and the wants that follow from these and allow them to control their experiences.

Empowering others likewise means allowing people to control their experiences. We expect people to take initiative and be motivated to act on their own. This means that we expect people to select goals and take action to realize them.

It is deeply satisfying to control your perceptions well—to realize your desires.

Control mechanisms have been invented several times in history (James Watt's steam engine governor is one of the best known examples). But it was not until 1927 that the feedback loop concept of control was completely analyzed and graphically illustrated. Modern "purposeful machines" such as the cruise control in your car and thermostat/furnace in your home are based on this concept.

The principle of control was introduced to behavioral science by Norbert Wiener in *Cybernetics* (1948).

People interpreting Wiener's presentation, using their existing event-based framework, created the impression that control is a step by step process, internal to the organism. This allowed the incorporation of cybernetics into the basic linear cause-effect scheme. It also allowed the understanding of cybernetics and control theory to mean control of the organism's actions, a misunderstanding that is widespread among behavioral scientists to this day.

A framework that accurately describes how and what people control has been developed. It is now called *Perceptual control Theory (PCT)*. This framework is laid out in the book *Behavior: The Control of Perception* (Powers, 1973). (see pages 81-82)

This book presents an engineering-oriented understanding of control, based on knowledge of neurology and physics, from a point of view inside the organism. A small group of researchers have built on Powers' work with his cooperation. A few behavioral scientists have tried to simplify the ideas of Powers' control theory and make them their own without understanding them properly. As a consequence, some popular literature perpetuates the misunderstandings of the 1940's and talks about control theory as about control of behavior.

Please note: Learning PCT is not learning a new way to behave. PCT offers a new and deeper understanding of how people have always behaved.

A person can be illustrated in summary form as one control system, as shown in exhibit 3.

The benefits you may derive from studying control include:

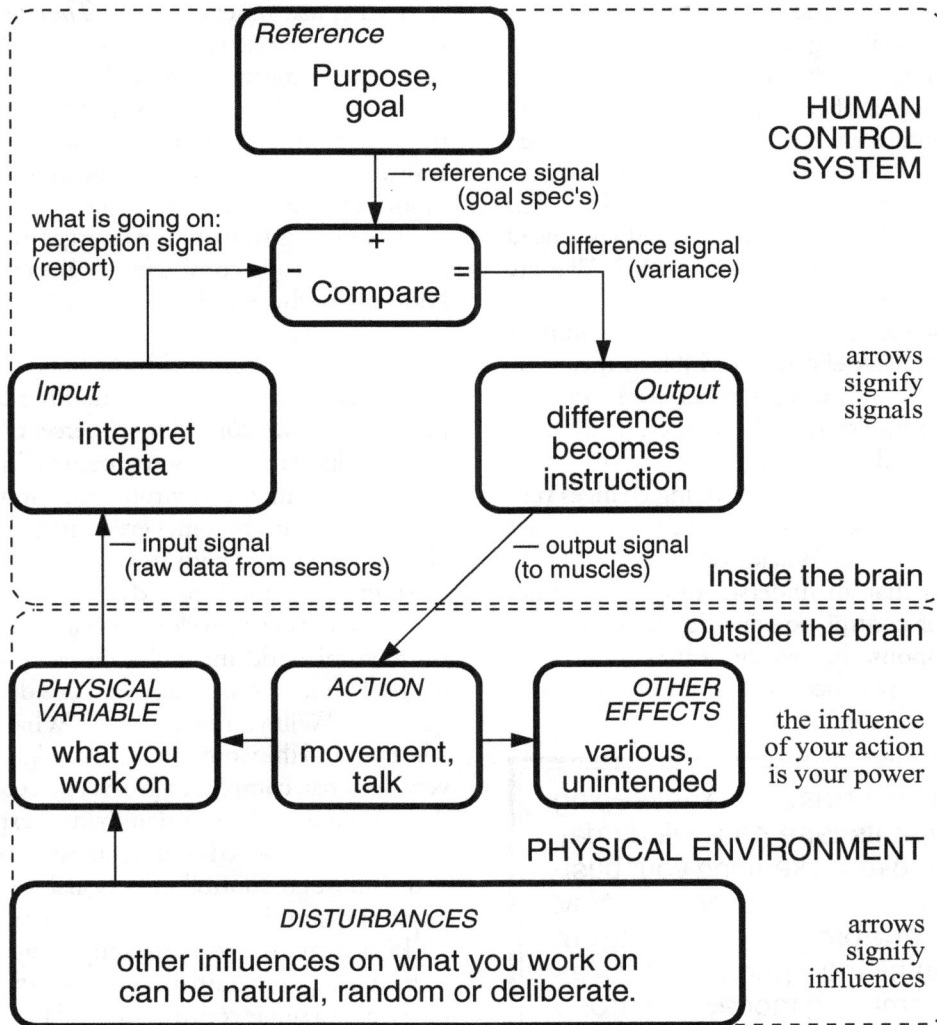

Exhibit 3. A person as one control system.

1. The realization that behavior is control of perception.
2. An understanding of how control works, including:
   - what is being controlled (perception, not output or action).
   - that control systems resist disturbances (other influences) acting on that which they are controlling. Consider how you resist praise ("Oh, gosh, you are exaggerating"), or criticism (in a performance review). Each disturbs your sense of self and you do what you can to counter the disturbance.
   - how control requires power—the application of resources to convert output signals into action.
   - the continuous, concurrent nature of activity in all functional parts of an automatic control loop.
   - what determines the speed of response to changing goals.
   - what determines the sensitivity to differences from the specified goal.
3. An understanding of what control looks like, such as:
   - that the action of resisting can be visible to an outside observer, but of no significance and even invisible to the control system. Have you ever leaned into the wind, raised your voice when you don't feel understood, or tossed your head when your hair obstructs your vision? All quite automatically.
   - that the action of resisting may have any number of incidental side effects, also visible to an outside observer, but of no significance to the control system. Examples might be noise created, energy expended and objects displaced.
   - that a variable you are controlling changes very little if it is well controlled.—It does not attract the attention of an outside observer.
4. Recognition that an understanding of the phenomenon of control explains the appearance of stimulus-response in behaviorism as well as the appearance of plan-action (control of output) in cognitive psychology.

> "Control can be as simple as keeping your car in its lane on a windy day or as complex as keeping your business profitable in a shifting economy. Control is the process of carrying out and maintaining a purpose, such as driving or making money. Control is purposeful behavior."

Controlling perceptions: Engineers commonly talk about control systems as controlling output, where output is understood to be a physical quantity subject to control. But think for a moment: If engineers at NASA are controlling the position of a satellite, are they really controlling the actual, physical position of the satellite? All they know about the position of the satellite is the reading of an instrument that tells them where the satellite is. This is their perception of the satellite's position. This is what they control. They compare the reported instrument reading with a desired instrument reading. Any difference suggests corrective action. If the reading is wrong because the instrument is out of calibration, the engineer will never know, unless some other instrument reading begins to show that something is radically wrong. All the engineer knows about the position of the satellite is what her instrument tells her. *The only thing she can possibly control is her perception* of the satellite position. Action is the means we use to influence the quantity we perceive. What stays constant is the perception; the action varies. It becomes clear that *all* control is control of perception, where perception is the sensed input signal in the control system.

The same is true of human organisms. The ends are controlled and consistent, the means vary. All we know about the world outside our brains (including our body functions) is what we sense as our perceptions. The view from inside the brain is the one that counts, and it is made up of nothing but our own perceptions. We control these perceptions.

The distinction between controlling quantities and actions (in the environment) and perception (inside the brain) becomes even more important and obvious as we deal with more complex perceptions like human relationships and conceptual understandings, where there is no clear correspondence between the physical world and highly developed perceptions in the brain. As an example, consider the loaded question: "Will you marry me?" What is marriage? The man and the woman each has a highly developed, very personal, complex high-level perception of what the word means, based on individual experience. As humans we do indeed control our sense of being married. But we are controlling personal perceptions, not physical quantities in our environment.

PCT demonstrates with compelling evidence that human beings function like one control system when focused on a single control task, and that they control their perceptions.

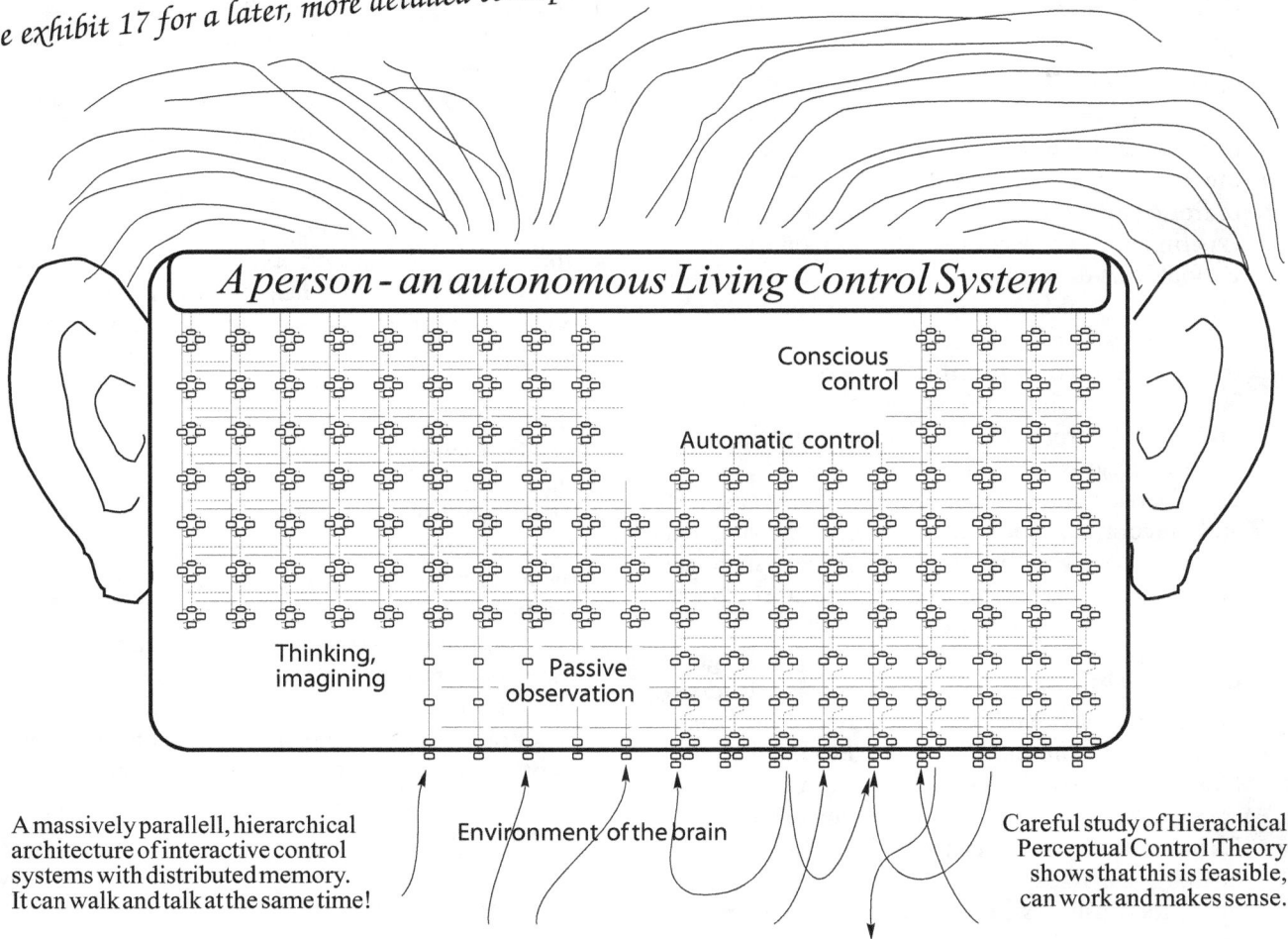

See exhibit 17 for a later, more detailed concept sketch.

A person - an autonomous Living Control System

Conscious control

Automatic control

Thinking, imagining

Passive observation

A massively parallell, hierarchical architecture of interactive control systems with distributed memory. It can walk and talk at the same time!

Environment of the brain

Careful study of Hierachical Perceptual Control Theory shows that this is feasible, can work and makes sense.

Exhibit 4. A person as a hierarchy of interacting control systems.

A person as a control system hierarchy

An extension of PCT called *Hierarchical Perceptual control Theory (HPCT)* suggests a construct of a human as a system of control systems. (Exhibit 4).

HPCT expands our understanding of control and we see

1. how control systems can form responsive and stable hierarchies, capable of controlling multiple variables at the same time, including:
   - how lower-level control systems get their goals from higher-level control systems, singly or in combinations.
   - how lower-level control systems can pass their perceptual signals to higher-level perceptual functions, singly or in combinations;

2. that it is "control all the way down." Control is not an occasional activity but a pervasive, multidimensional, continuous fact of how our nervous system operates.

While exhibit 1 is a metaphorical way to portray an organization, exhibit 4 (and 17) is intended as a functional description of the organization of the human nervous system.

HPCT suggests elegant explanations of how we 1) think, imagine and dream; 2) observe passively; 3) control many functions without conscious attention; and 4) control selected perceptions with full attention.

Some evidence for this hierarchical organization of control in humans is easy to demonstrate for the lowest 4-5 levels of vision and muscle coordination.

Demonstration of a human hierarchy

The following illustration of two levels of control is adapted from Robertson and Powers (1990, p. 21)

To demonstrate several "nested" control systems in the body, begin with one which is exemplified in the spinal reflex loop. A subject (S) extends an arm in front, with instructions to hold it steady, and the experimenter (E) places a hand lightly on top of S's. E should make sure that S is not holding the arm limp. E then gives a sudden sharp downward push, and S's arm appears to rebound as if on a spring. An electromyograph verifies that this is an active, innervated correction, not simply muscle elasticity. The initial position of S's arm makes no difference, and the initial muscle tensions involved also make no difference. S can be asked to hold the arm in a different position, and the control action will be the same, showing that the reference signal [want, goal] for the system can be altered and the system will continue to correct its action to the new reference setting.

Higher-level systems derive their feedback signals from sets of lower-level feedback signals. To demonstrate the next level, E now instructs S to extend the hand as before and E again places a hand on top. Now E tells S to swing the arm downward as rapidly as possible, as soon as S feels E's downward push. E's hand must be in contact with S's to make the push as sharp and unexpected as possible. Immediately upon the push, S's lower-level systems return the arm to its initial position, because they act within the latent period of the higher-level feedback signal. The initial correction is nearly completed before the higher-level resets the reference signal.

Your timing observations will agree with engineering requirements for stability in hierarchical systems of control systems. Individual introspection demonstrates the evidence for higher levels. After all, the individual is the only one who has access to the internal workings of the individual brain. Other evidence and illustrations of hierarchical control will be presented in the next article in this series.

Insight

Once you accept the concept of people as autonomous living control system having a system of internal understandings and purposes and controlling their own perceptions, you gain a different outlook.

- You see conflict, cooperation, motivation and commitment in terms of purposes and perceptions.

- You recognize that a person's actions are of secondary interest. They are the means to achieve a desired perception and as such are quite incidental. You recognize that you cannot tell what a person is really doing (controlling what perception) by watching what the person is doing (action).
- You recognize internal conflict—within the individual—as a major source of problems.
- You see how you can support others and help them resolve their conflicts by helping them review their purposes and perceptions, *not* their actions. This insight is in direct contradiction to conventional wisdom, which holds that the only thing that is clear and tangible is the actions. It is hard to let go of a focus on the actions we see and experience—and deal instead with the perceptions and wants that drive those actions.
- You understand the importance of acknowledging the other person as an autonomous living control system.
- You learn how a leader can find out what the associate's (subordinate's, peer's, superior's, prospect's, friend's) goals and perceptions are, and work to encourage him to align his understanding and wants with the goals of his organization. This alignment is at the heart of good sales, negotiating, management, and hiring practices.

Implications for leadership

Telling people what to do, complete with action plans, implies linear cause-effect thinking. It does not work very well in a world of changing circumstances. It is better to specify the desired outcome and leave the momentary action to the capable subordinate. Hierarchical control systems work this way with surprising agility.

Organizational development requires an understanding of individuals, just like algebra and calculus require an understanding of arithmetic. Arithmetic is rigorously proven, based on a few postulates. Scientists are confident when they apply algebra and calculus to problems.

Scientific understanding of individuals is limited and far from proven. Managers have little confidence in their design of teams and organizations. They experiment to find what works by trial and error—with mixed results.

With a clear, testable and proven concept of one individual as a living, hierarchical control system, managers can design teams and organizations based on a valid conceptual understanding.

Some observations:

- As your understanding of the individual changes, old conclusions about organizations must be reviewed in light of the new knowledge.
- When Dr. Deming said: "..the aim of the system must be clear to everyone in the system..," he appeared to intuitively recognize that people can perform isolated functions in an organization, but that each is a comprehensive, autonomous control system.
- Each individual adopts personal purposes and ways of perceiving, based on personal experiences.
- It is impossible to "dial" a want into another person in the same way that you set a thermostat. In other words: You cannot command a person what to want. A person must voluntarily adopt an understanding and the wants that go with it.
- All of the active "organization goals" reside in the heads of individuals, not in printed statements.

Giving and taking orders

Let us review, from a PCT perspective, what happens when one person gives an order to another person. The person giving the order has to imagine something she wishes to perceive. She has to translate this mental perception into words or images that are emitted into the physical world in the form of oral instructions, memos, diagrams, or gestures.

After this the order leaves the giver's immediate control. The physical light, sound, or touch impinges on the order taker's senses. He hears the sounds, sees the words and images, and senses the touch. He has to turn what he senses into meaning. He has to translate the physical phenomena he experiences into a coherent perception, which he believes will satisfy either the spirit or the letter of what the order giver intended.

Where does he find meaning that can be matched to the sounds and images he perceives? It can only come from his own memories of experience. The order he obeys is one he puts together from perceptions recalled from his own memories—an internally constructed perception he will try to match by acting on the world.

This link from one person to another is complex and loose, even under the best of circumstances, with the best intentions of cooperation. The manager should not be surprised if things don't turn out quite the way that he or she intended.

An understanding of PCT helps both the order giver and the order taker focus their attention on this difficult process in a way that makes it as effective and satisfying as it can be.

Summary

An understanding of PCT gives insight that shows you how to align personal goals among associates with those of the organization. An organization can function as a cooperative, responsive, productive entity where it is a pleasure to work.

This discussion of organizations and their dependence on psychology for management practice has introduced PCT as an explanation that describes the way humans actually work. Some of the lessons PCT holds for effective human interaction have been shown.

References

Cziko, Gary A., "Purposeful Behavior as the Control of Perception: Implications for Educational Research." *Educational Researcher, 21*:9, (November 1992), pp. 10-18, & 25-27.

Ford, Edward E., *Freedom From Stress.* Scottsdale AZ: Brandt Publishing (1989).

Gibbons, Hugh., *The Death of Jeffrey Stapleton: Exploring the Way Lawyers Think.* Concord, NH: Franklin Pierce Law Center (1990).

Marken, Richard S. (Ed.). (1990). Purposeful Behavior: The Control Theory Approach. *American Behavioral Scientist,* 34(1). Thousand Oaks, CA: Sage Publications.

Marken, Richard S., *Mind Readings: Experimental Studies of Purpose.* New Canaan, CT: Benchmark Publications Inc. (1992).

McClelland, Kent (1994), "Perceptual control and Social Power." *Sociological Perspectives,* 37(4):461-496.

McPhail, Clark., *The Myth of the Madding Crowd.* New York: Aldine de Gruyter (1990).

McPhail, Clark., William T. Powers, and Charles W. Tucker, "Simulating individual and collective action In temporary gatherings." *Social Science Computer Review, 10*:1, (1992) pp. 1-28.

Petrie, Hugh G., *The Dilemma of Enquiry and Learning.* Chicago: University of Chicago Press (1981).

Powers, William T., *Behavior: The Control of Perception.* CT: Benchmark Publications Inc. (1973, 2005).

Powers, William T., *Living Control Systems: Selected Papers.* New Canaan, CT: Benchmark Publications Inc. (1989).

Powers, William T., *Living Control Systems, Volume II: Selected Papers.* New Canaan, CT: Benchmark Publications Inc. (1992).

Richardson, George P., *Feedback Thought in Social Science and Systems Theory.* Philadelphia: University of Pennsylvania Press (1991).

Robertson, Richard J. and William T. Powers (eds.), *Introduction to Modern Psychology: The Control Theory View.* New Canaan, CT: Benchmark Publications Inc. (1990).

Runkel, Philip J., *Casting Nets and Testing Specimens: Two Grand Methods of Psychology* (1990, 2007) Menlo Park: Living Control Systems Publishing.

Soldani, Jim, "Effective Personnel Management: An Application of Perceptual control Theory." Posted at www.livingcontrolsystems.com

Wiener, Norbert, *Cybernetics or Control and Communication in the Animal and the Machine.* New York: John Wiley (1948).

Perceptual Control — Management Insight for Problem Solving

Number two in a series of three articles on PCT
An early version of this article appeared in
Engineering Management Journal Vol. 6 No.3 Sept 1994

Perceptual control — management insight for problem solving

ABSTRACT

This article suggests that managers focus on the wants and perceptions of their associates instead of their behavior in a questioning approach to problem solving. This recommendation is based on the first successful, demonstrably valid concept of the basic operation and structure of our nervous system. A discussion of the nature of the concept, a do-it-yourself demonstration, and detailed instructions on how to solve problems are included.

INTRODUCTION

This article applies Perceptual Control Theory (PCT) and Hierarchical PCT (HPCT), introduced in the first article, to problem solving situations. The architecture presented in exhibit 4, (page 21), is a representation in principle of Hierarchical Perceptual Control Theory. The idea of a person as a hierarchical system of control systems seems both preposterous and incomprehensible unless some of the underlying principles are understood. Out of context, the demonstration in this article of a person acting as one control system may be dismissed as a curiosity. If so, conflict resolution by means of mapping and influencing wants and perceptions becomes just another unfounded prescription for action. The purpose of this article is not to provide an exhaustive technical description of HPCT, but to explain conflict resolution. I shall limit the technical content to a few comments about the nature of the theory on which the recommendations are based.

Focusing on one of the many control systems active within the hierarchy, you can perform a do-it-yourself demonstration with a friend. This will show you how invisible control is (because people have never learned to recognize it) and provide an "A Ha" experience for both of you. You can illustrate

> It is not necessary to understand.., because people *are* control systems and control whether they understand it or not. But if you *do* understand, you can solve problems more effectively.

conflict and cooperation. From an understanding of control and conflict follows insight into organizational interaction and lessons about how conflict can be resolved.

The careful student will find this a fully integrated, useful explanation of how thoughts (perceptions and purposes) become actions, results and feelings. It has much to say about how we grow up, live our lives, interact, and manage organizations effectively.

Understanding the nature of HPCT

If nature had evolved Personal Computers, a society of non-technical people would most likely suggest that computers are too complicated to understand. A non-technical scientist researching how computers function would press keys on the keyboard, observe what happens on the screen or with the printer, and try to make sense of it through many experiments, using statistics if results were inconsistent. It would be extraordinarily difficult to learn anything about the internal organization of the computer that way.

To understand and reverse-engineer a computer, it would be necessary to

a) Understand the physical sciences.
b) Make a lucky guess about the nature and structure of the computer's various parts.
c) Test the resulting functional model against the function of an actual computer.

With such an understanding, borne out by successful tests, the user could do more with the programs that run on the computer, change some of the programs, and thus could accomplish far more than other computer users.

We have been told for centuries that the human brain is too complicated to understand. Research results have been so inconsistent that statistics are employed to indicate the validity of observations made.

Perceptual Control Theory successfully attempts to reverse-engineer our nervous system, create functional models to simulate it, demonstrate that the basic concept is valid, and point the way to more effective research methods.

Levels of perception and control

The *vertical dimension* in exhibit 4 is "Levels of perception." Starting from the bottom, a low-level input—a neural current created by a nerve ending "tickled" by some physical phenomenon in the environment, such as light falling on a single cell in the retina—is combined with other inputs, creating a perception signal at a higher level, in turn combined to create a signal at a still higher level. At the higher levels, a branch of the perceptual signal can be stored in memory and later played back as a reference signal. (HPCT incorporates distributed memory to explain imagination, automatic control and passive observation).

Perception and control starts with intensity signals from neuron sensors and develops successively higher level perceptions, presently thought to be intensity, sensation, configuration, transition, event, relationship, category (language fits here), sequence, program, principle and, at the highest level, systems concept (the way the world is). Each successively higher level of perception builds on the immediately lower ones.

The *horizontal dimension* is "Examples of perception." At the lowest levels, we perceive light, vibration, pressure, temperature, joint angles, tendon stretch, smell, taste and physiology (which we sense as a part of feelings). At higher levels, we form perceptions of things and concepts like clothing, food, personal relationships, honesty and employment. These principles and system concepts are descriptions, explanations and mental models of the world, in many areas of knowledge, which we learn and decide to believe in. Taken together, they constitute what we call cul-

ture, science, religion, ethnicity and so forth. The insight HPCT offers is that these principles and systems concepts are perceptions in themselves. In daily language we talk about understanding, belief, or generally "the way the world is or should be."

Based on the *systems concepts* we have internalized, in comparison with the world as we see it, we select *principles* to live by: priorities, values, standards. These in turn—again in comparison with perceptions of the current world, determine the *programs* or action plans we carry out. From these follow *sequences,* or methods made up of *events,* work elements needed to carry out the programs we have chosen. Events require control of muscles and body chemistry at the lowest levels.

Validation of HPCT is found in numerous experiments and in the development of infants (Rijt-Plooij, 1992, 2010). The Plooij's have identified 10 highly predictable periods of mother-infant crisis in the first 18 months of life. They have found that the newborn infant controls at the second level, with perception of configuration emerging at 7-8 weeks, perception of transitions at 11-12 weeks, events at 17 weeks, and so forth. The principle level emerges at 14.5 months and the systems concept level (including the notion of self) towards 18 months.

> Your action illustrates plainly the phenomenon of perceptual control—we act in opposition to disturbances to develop and maintain perceptions we want.

With this brief outline, I hope you can see how your own perceptions "behave" from your highest systems concepts level down to the lowest levels. You do not have to have a detailed outline of HPCT to realize that what you really want—what is important to you—you make come true as best you can. We control our world as we perceive it from the time life began until we die.

The demonstration that follows shows how you can focus your attention on control of something, in this case a single visual relationship, and how your mind makes it come true, working through your hierarchy of control systems, physiology and muscle fibers.

A demonstration of control

You can perform a practical demonstration, wherever you are, with the simple prop of two rubber bands joined by a knot. Just get a friend to help you play a game. This game will illustrate all the elements of human control, their interactions and functional relationships. This description follows Runkel (2007, pp. 103-106).

I am hopeful that placing this demonstration in the context of the larger hierarchy gives it more meaning in your mind. When dealing with every aspect of your own life—requesting water instead of juice to drink; insisting on telling the truth because that is honorable, for example—you are specifying and controlling complex perceptual variables, just like you or your friend control a rather simple one in this demonstration. The rubber band is a very simple environment, where disturbance and action have a direct, obvious influence on the variable.

This rubber band demonstration becomes a functional representation of how we live our lives. The visual relationship you select represents anything you want at the moment, and this variable, as you perceive it, instant by instant, represents your perception of the world, corresponding to your want.

Join two rubber bands by a knot. You hook a finger into the end of one rubber band and your friend hooks a finger into the other (Exhibit 5).

Exhibit 5. The rubber band demonstration.

Tell your friend something like: "You are the experimenter. Move your finger as you like. Watch what I do. When you can *explain* what is *causing* me to do what I do, let me know."

When you sit down with your friend, place yourself so that the knot joining the rubber bands lies above some mark you can see but which your friend probably will not notice—a small mark on a table top or paper, a piece of lint on your knee. As your friend's finger moves, move yours so that the knot remains stationary over the mark.

By deciding to keep the knot over a target, you have adopted a particular visual relationship as your want. When something disturbs this relationship, you will restore it. You will move in any way necessary to do that.

Of course, you cannot keep the knot stationary if your friend moves faster than you can act. Some people playing this game seem to want to move abruptly, too fast. If that happens, ask your friend to slow down. The lessons to be learned will be much more obvious to both of you if you are able to keep the knot continuously over the mark. You might say: "Don't move so fast. I can't keep up with you."

Your friend will soon notice that every motion of her finger is reflected exactly by a motion of yours. When she pulls back, you pull back. When she moves inward, you move inward. When she circles to her left, you circle to your left. You must do that, of course, to keep the knot stationary in this particular environment. Your action illustrates plainly the phenomenon of perceptual control—we act in opposition to disturbances to develop and maintain perceptions we want.

Notice that you perform many different acts to maintain your perception of the visual relationship. You move your finger to the left, to the right, forward, backward, and diagonally at varying speeds.

Most people, when they announce that they can explain what is causing you to do what you do, will say that you are simply mirroring what they do, or imitating it, or words to that effect. Some will put it more forcefully: that whatever they do, you are acting in opposition to it. Almost all will say or imply that *they* are the *cause* of your behavior.

A few people will notice that the knot remains stationary. That is an excellent observation, but not quite an explanation of cause. Agree, but keep asking: "What is *causing* me to do what I do?" Most people will say that your intent is to do something in reaction to them. But then you deny that. They will eventually give up and ask: *"All right, what is causing your behavior?"* You explain that you have been keeping the knot as close to the mark as possible, and that any *difference* caused you to do what you did.

You moved to oppose any motion of the knot away from the mark, not to oppose her. Your motivation had nothing to do with what your friend might have been trying to do; you did not care. *You watched only the knot and the mark.* Indeed, if you had not been able to see your friend's moves, your actions would have been identical. Watching the knot and the mark carefully, you cannot pay close attention to her movements at the same time. There is no need to.

Reactions of experimenters will vary widely. A few will accuse you of being devious and go away grumbling. Most will be surprised, even dumbfounded, to have missed the obvious. A few will find many of their previous ideas so shaken that they will think about it for days afterward.

Play the game with your friend. Play it with several friends! This game is an important part of this introduction. It only takes a few minutes. Please be sure to actually do the demonstration with another person. If you just visualize it, you will miss the insight of just how invisible the phenomenon of control is.

Suppose you played this game with 10 of your friends. Let us say that one was in fact able to explain (without coaching) that you were only holding the knot steady over the mark and acted to keep it there. That still means that 9 out of 10 failed to recognize the phenomenon of control when it was right in front of them. They have never been shown what control is or how to recognize it. Without an understanding of control, they are literally blind to a phenomenon that is fundamental for all living organisms.

Repeat the game with visibility for both of you. This time you are the experimenter. When your friend has seen the simple explanation: that the action is a function of the experimental setup—the rubber bands—and follows from her want to keep the knot over a mark, ask your friend to do it once more and use a pen to trace the action.

Exhibit 6 shows what the trace might look like. Notice that the knot moves a bit, erratically, about the target. If you think of this as a production process, this movement might represent variability of production quality. The slower you perform this demonstration, the better quality you can achieve, because your control will be better.

Now we focus on your friend's visible behavior and ask: "What can a reasonable observer conclude about your friend, based on what the observer can see of your friend's behavior?" What is your answer? Now that you have acted out this demonstration and considered the question, would you agree that you cannot draw any conclusions about your friend from her behavior? Your friend's behavior is clearly a product of what your friend wants (a visual relationship, specified in her mind), combined with the disturbances (your pulling on your band) acting on what she is controlling (her current perception of the visual relationship) . Her behaviors are what they have to be under the circumstances, given all the functional elements, their influences and interactions.

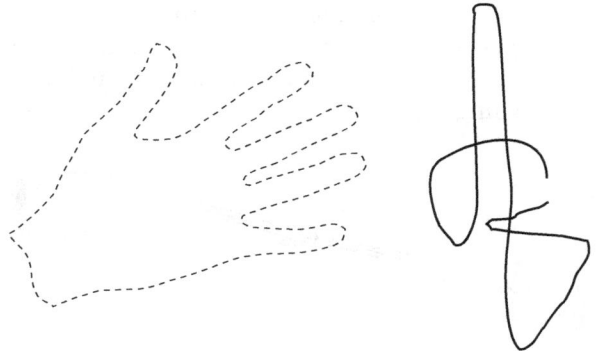

Exhibit 7. Only muscle action is visible to an outsider.

Exhibit 7 suggests that the only part of everyday behavior an observer can see is the action. Hidden from view by the hand are: 1) your friend's want, 2) the disturbing influence the experimenter has on what your friend wants, and 3) many aspects of the environment. What your friends and associates want at any moment, how they perceive it, and what disturbs it is seldom visible to an observer.

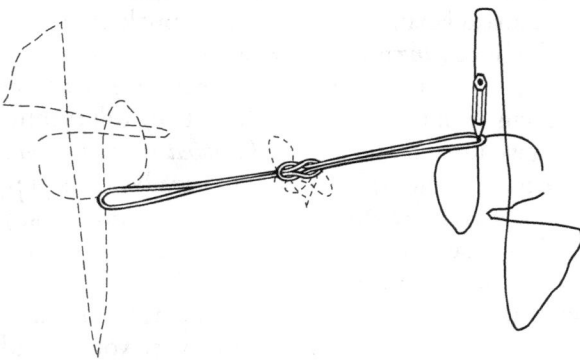

Exhibit 6. Tracing the rubber band action.

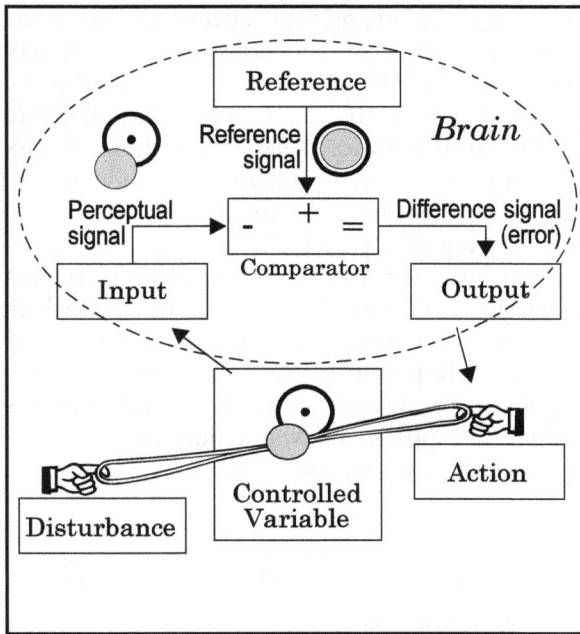

Exhibit 8. Rubber band diagram.

Compare with exhibit 12. Taken from a classroom illustration, this diagram shows a ping-pong ball over the knot, making it more visible.

This demonstration and the diagram in exhibit 8 clearly illustrate wants (reference signals) and perceptual signals, the difference between them, output signals that provide instructions for action, the actions themselves, which influence the variable we control, and other influences (disturbances) on the variable.

This demonstration is more easily appreciated when you can be face-to-face with the person doing it, talk about it, see the diagram as it unfolds and ask for clarification. Notice that everything is apparent. You are able to see, question, and discuss the elements and their relationships.

This is a simple but complete way to understand what is going on. The control model provides complete diagnostic tools for any interaction between people—whether in cooperation or conflict.

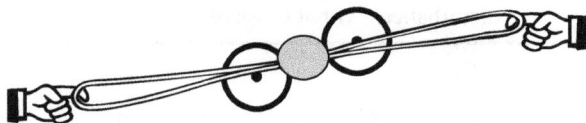

Exhibit 9. Rubber band illustration of conflict.

Conflict

Repeat the experiment with your friend, but this time with both of you controlling your own visual relationship (Exhibit 9). Your target is the one closer to you. The moment you start, you will both pull as far and hard as you dare (not wanting the rubber band to snap and hurt you) in your own direction. If you repeat the experiment with a rope instead of rubber bands, you will find that the stronger person can reach her target, while the weaker is frustrated. The waste of effort is obvious. Conflict can arise in other ways, for example if the two players perceive a single target differently, from different angles.

Cooperation

With a three-part rubber band and three players, you can demonstrate cooperation (Exhibit 10). Two players can both influence the knot with one agreed-upon target, with the third player providing a disturbance. The cooperating players can pull in different directions and with different forces (one can even slightly counter the other), in such a way that the net result compensates for the disturbance, or they can work completely in parallel to compensate with a minimum of total effort.

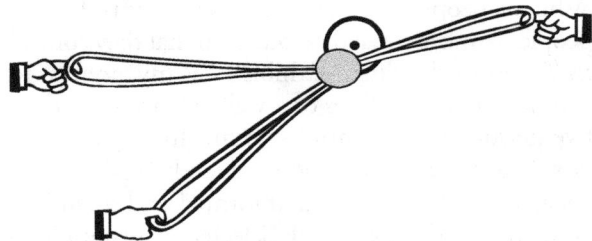

Exhibit 10. Rubber band illustration of cooperation.

An assertive person

The concept of assertiveness intuitively recognizes our nature as control systems. In exhibit 11, an assertive person claims the right to control his own perceptions in several different ways. If you claim these rights for yourself, how about granting them to others? That means recognizing your fellow man as a living control system, just like yourself. Depending on just exactly what it is your fellow man understands and wants, you may be happy to work side by side, or want to put great distance between him and yourself. As shown in the demonstration of conflict and cooperation, what we want and how we look at the world do make a difference.

I claim the right to:

- *be treated with respect*
- *have and express my own feelings and opinions*
- *be listened to and taken seriously*
- *set my own priorities*
- *ask for what I want*
- *get what I pay for*
- *make mistakes*
- *assert myself even though I may inconvenience others*
- *choose not to assert myself*

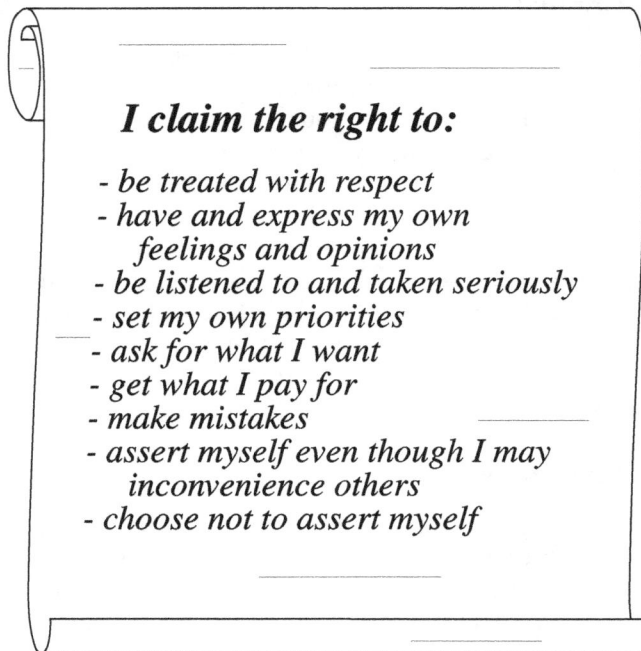

Exhibit 11. An assertiveness bill of rights (Zuka, 1983).

An effective person

While all people are equal in that we all control, some people are more equal than others in that they control *well* most of the time. Exhibit 12 is my attempt to summarize the qualities of a well balanced, productive person as one control system. In the reference box, I have shown the concept of levels of perception collapsed to the statement: Informed understanding → Considered priorities → Selected wants, indicating that a person's wants (right now, in relation to present circumstances), are not selected at random in a vacuum, but derive from higher understanding. The wording in the other boxes must also be read as a composite of the capabilities of the entire hierarchy. My point in offering exhibit 12 is to suggest that a person who is cool, calm and collected in most circumstances, is a pleasure to deal with and very productive, can properly be portrayed this way—a very capable system of control systems.

This "portrait" allows for a great variety of wants and perceptions. It is easy to see how people can be labeled as having different "personalities," classified in popular books as "difficult people," and stereotyped as "dysfunctional." People develop different understandings, priorities, wants and ways of

perceiving/interpreting their experiences. The entire structure of perceptual functions and stored perceptions is our individually subjective *reality*. (See *We can never know REALITY*, page 63). Our ability to control our lives varies, depending on how effective this subjective *reality* is in helping us deal with the REAL world outside our minds.

Thinking of a person this way gives the manager obvious diagnostic tools: In any situation, ask questions about what the person wants (and which wants are more important), what the person does *not* want, how the person perceives the situation, including alternative interpretations, how satisfactory the comparison appears and what actions the person has considered in imagination.

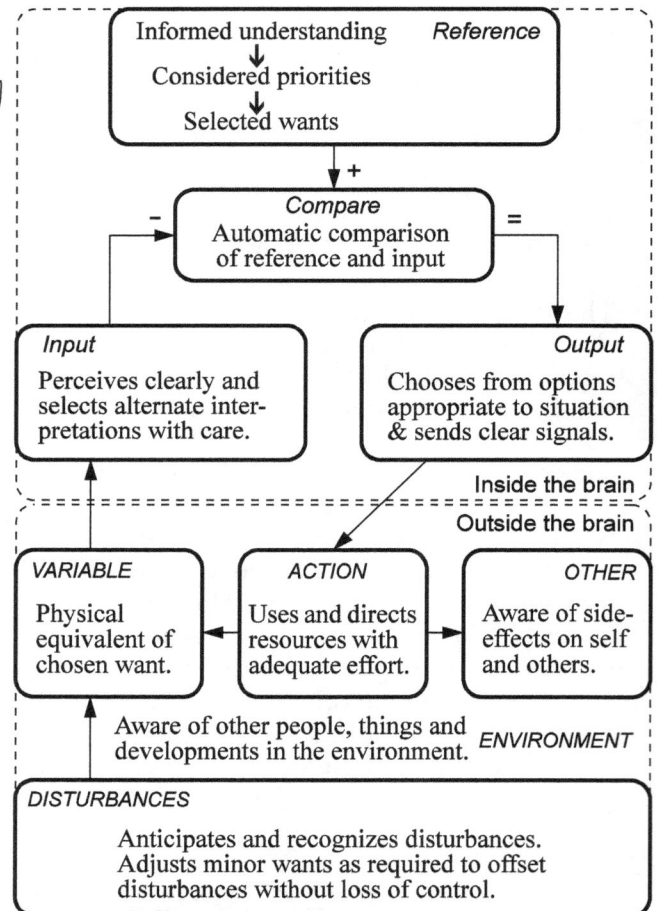

Exhibit 12. An effective person

People interacting

Exhibit 13 shows a framework for understanding the interaction between people, whether in conflict or cooperation. Here, two brains are shown, acting in a common environment (outside the body, of course). Each person is controlling a perception of some physical variable as that person wants to, by acting on it. If the chosen variables are related or even the same one (say the balance of a tandem bicycle), it quickly becomes obvious that a variable is subject not only to disturbances from the environment in general, such as crosswind, but also that each person's action becomes a disturbance to the other. Even side effects of independent actions become disturbances to the other. (The balance is affected/disturbed if one turns around to enjoy the view).

In this illustration, person #1 can represent your associate or a prospective customer. Person #2 can represent another associate or yourself or your prospect's associate. You can readily extend this illustration with Person #3 in another department, Person #4, #5 etc., all interacting in the same physical environment. Exhibit 13 provides the framework only; the boxes are not filled in with specific understandings, wants, perceptions, output options etc. Each person in exhibit 13 lives in a personal "world" of wants and perceptions. Besides personal variations, these worlds can be very different because of professional specialization, studies, experience, and responsibility.

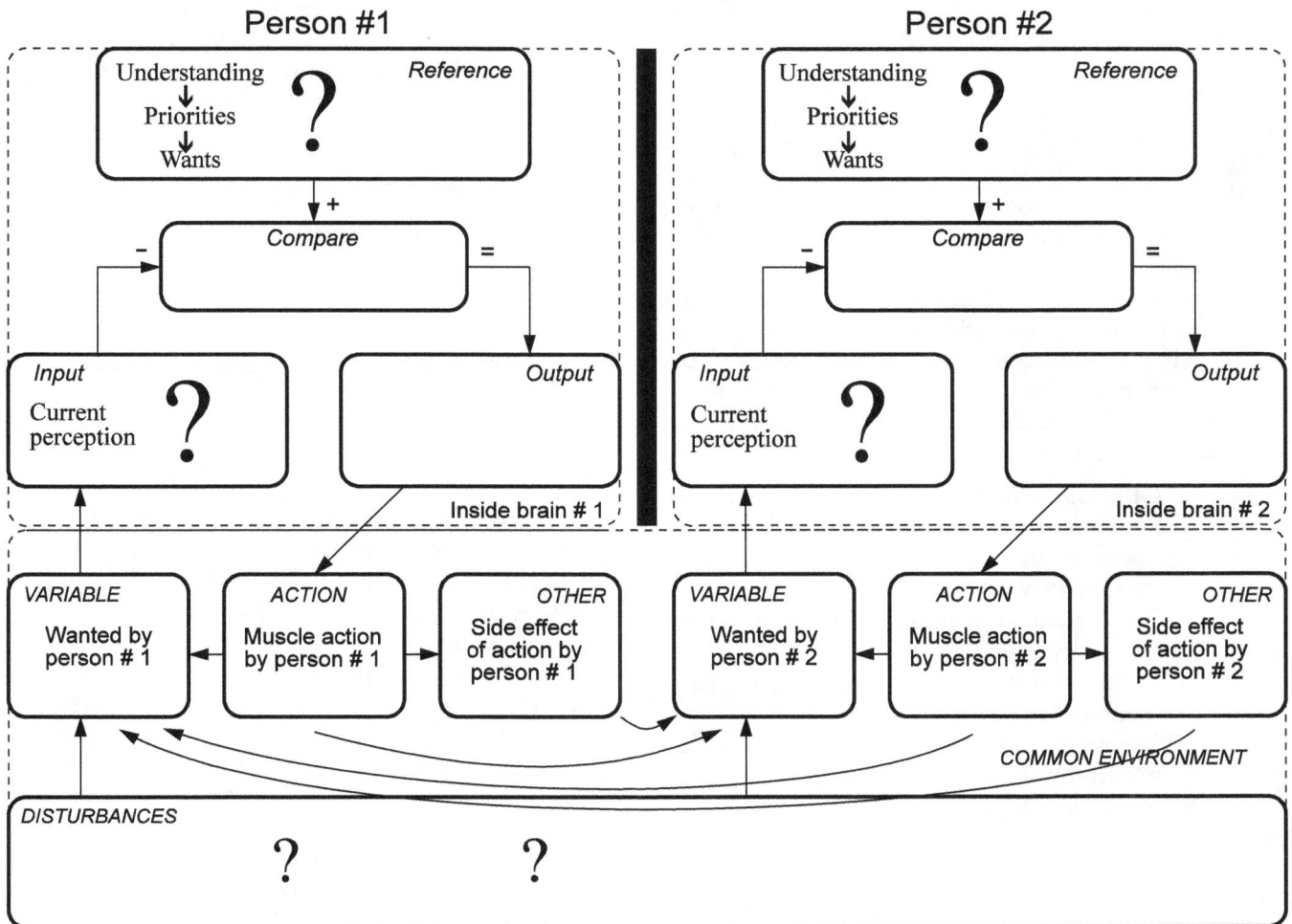

Exhibit 13. Two people interacting.

Organization

Exhibit 1 portrayed how we often think of a hierarchical organization and how we develop specialized goals for individuals in different parts of the organization. Note the visual similarity between that hierarchical goal structure and the hierarchical control structure shown in exhibit 4.

Exhibit 13 shows how, once those goals have been communicated and accepted, an entire company can more properly be portrayed as individuals working side by side in a common environment. Development, communication and agreement on goals is not easy. Telepathy between brains is not possible. (The black line represents a barrier between brains). Everyone must interact through the environment, as exemplified by the order giver and taker on page 23, even if the controlled perception is a high level mental construct such as "honesty," that has no physical equivalent in the environment.

Respect

Respect, ethics, morals—a sense of right and wrong— follow naturally from an understanding of HPCT. You realize that you are a living control system, and assert the right to control your own perceptions as freely as possible within the constraints of nature. In fairness, you accept that your fellow man deserves and asserts the same right.

If you want to not only "live and let live," but also want to support your fellow man, HPCT shows 1) what supports effective control, 2) what defeats it, and 3) what disturbs it.

1 a) Offer the best possible, validated, factual information for consideration. This helps your associates develop understanding and select appropriate wants.
  b) Allow them to perform freely and experience the results.
2. a) Misleading information can create unattainable wants and frustration.
  b) Too much help does not allow your associate to perform—to experience effective, satisfying control and to learn from it.
  c) Promises or threats distort purposes and can create conflict.

3. Judgements of, remarks about, and criticism of action/behavior focus on the incidental means, not the purposes and perceptions of a living control system. This does not help at all, but disturbs your friend and creates conflict. It is impossible to convey a sense of respect when focusing on the action/behavior of another.

Lessons for managers

It is not necessary to understand how control works to live, because people *are* control systems and control whether they understand it or not. But if you *do* understand, you can solve problems more effectively.

From the detailed insight of HPCT, managers can learn this most important insight: *Judging action/ behavior is next to useless.* It tends to cause conflict, not solve it. Wants and perceptions are what should be discussed so they can be reconsidered. When they change, action changes automatically. The interactions (horizontal arrows in the environment) portrayed in exhibit 13 change when the wants and perceptions of either or all parties change.

Mapping and influencing wants and perceptions

We have seen how exhibit 13 represents people working side by side; brains living in separate personal "worlds" of wants and perceptions in a common environment. We understand that actions are the result of an automatic comparison between current perceptions and related wants. Exhibit 13 shows clearly that if we want to understand (and influence) the actions of others, we must "map" the blank spaces in the areas of wants and perceptions. We know that people have wants and do perceive. The question is: What are the wants? What understanding and priorities are they based on? How does the person interpret inputs?

By *mapping and influencing wants and perceptions* you can explore the unknown territory of other minds. Ask questions. Where a person is unclear, your questions help her consider where her wants come from and alternate ways of perceiving a situation. You can ask what actions she has considered in her imagination, and what she thinks the results would be. *Mapping* can range from gentle exploration to challenging questions which help her consider and revise her wants and perceptions as they are being mapped. You can use the mapping approach in a professional discussion, in a sales presentation, in a performance coaching review, and in a firm but non-judgemental discipline session. (This will be discussed more in the third article).

The result of mapping is self- and mutual understanding. Every person involved in a cooperative task will clearly understand the relationships between their various wants, perceptions, actions, results and side effects. They can work things out and support each other.

Mapping can involve a whole team. Let me show how you can facilitate a simple conflict resolution between two people. A male associate may ask for your assistance in order to resolve some problem, or you, as his manager, peer or friend may approach him. You can work one-on-one with him alone. Your questions help him think through both sides of his conflict and draw his own conclusions.

A basic methodology might be as follows:

1) He asks for a meeting (or you do).
2) Ask him what happened and concerns with the other.
3) Ask him about his own wants in relation to the other.
4) Ask him what he thinks the other's wants, perceptions, and possible choice of actions are.
5) Ask him to compare. Does he see any conflict between his own wants and perceptions and those of the other as he understood them (or you clarified them)?
6) If yes, ask him if he wants to commit to work on a way to resolve the conflict.
7) If yes, coach and support him as he develops a plan to change wants, perceptions, capabilities and the environment to eliminate the conflict.

The point of this approach is to ask about goals and any conflicting goals and ask him to consider outcomes of his different options until he decides on a course of action that is best for him in the context of his agreement and capability to support the organization. You can renegotiate if you represent "the other" and support as appropriate.

Things to avoid when asking him to map himself and others:

- Do not dwell on the action that may be the reason for mapping. At no time do you criticize him. You conduct the entire session by asking questions, offering advice only when it is welcome.
- Do not *ever* tell him what you think, but ask if he would like to have information when you have something relevant to say. If you impose your opinion on him, he perceives your message as an attempt to control him and he will resist. He is concerned about what he wants, not about what you are saying.
- Do not dwell on his feelings, (it is not productive) but ask about what *causes* them, namely his goals and how they compare with his perceptions. (That gives him a way to deal with his feelings).
- Do not take over his responsibilities and try to do his thinking for him. Living control systems must do their own thinking in order to function effectively. Your role is only to ask questions (and teach when asked).
- Do not ask him why he has behaved in a certain way. He must now defend ineffective choices in the past.
- Do not bring up a negative incident from the past. It is beyond his control at this point.

As you explore the things he wants, you are not limited to things he mentions. As an experienced person, you can ask about related wants or reasons for these wants. For instance, if he has an internal conflict—incompatible wants—you can ask him about his priorities, which will help him to resolve his conflict. If he does not tell you what he wants, you can employ "the test." You guess what he wants, then disturb it and watch to see if he resists the disturbance consistently. (Runkel 2007, p. 115 & 150).

This approach is not soft and wishy-washy, leaving everything up to your associates, and you powerless. You will find that the approach outlined here is more effective than telling people what goals to adopt, assuming that they do, and talking to them about what they *do*—their actions.

Through careful and persistent questioning, you help your associates focus their attention on issues (you can raise issues related to company goals) that are important to them and help their mind to come up with solutions to what they agree are their problems. Over time, you become their trusted friend, someone who cares.

How is this different?

Conventional psychology teaches us that the only thing we can legitimately study and deal with is peoples' behavior. It is widely understood that the purpose of conventional psychology is the prediction and control of behavior. This behavioristic point of view encourages managers to think of people as something to be manipulated. What can we do to get our people to be the way we want them to be? How can we motivate them? How can we get them to come on time, work harder, show more loyalty to the company, pay more attention? In short: How can we control their behavior?

When we are unhappy with the results of the performance of another, we ask: Why did you do that? Can't you do something better? We tell people: You can't do that; your behavior is unacceptable! Here is what I would do if I were you... This is the accepted method in that situation. If you say this..., the customer will do that... We focus on and try to reinforce, reward, train and modify *behavior*.

The questions above often lead to defensive excuses, conflict and resentment. Only accidentally may they lead to a productive discussion of wants. It does not make matters easier that the term behavior itself is poorly defined and confusing. Behavior refers to action, but is invariably defined by the result: harassing behavior, loving behavior, cooperative behavior, leadership behavior, etc..

HPCT explains how we develop our own understanding, make our own choices based on our values and standards, and act freely to control our own perceptions. The last thing we want is for someone else to control our behavior.

PCT shows that *action is a normal by-product* of wants, perceptions and circumstances. When we are unhappy with the results of the performance of another, it is best to ignore the action/behavior—the by-product or symptom—and ask instead about wants, perceptions, and disturbances, which are the causes. (Exhibit 13).

You stimulate creative thought through questions rather than manipulative coercion. Respect for your associates' internal world of wants and perceptions is critical.

When you change from trying to control your associates' behavior to asking them to deal with both their own and their organization's wants and perceptions, your associates learn to think, sort out internal conflict, and develop effective plans. You allow your associates to control well: to satisfy personal and company wants at the same time. You are seen as a leader and teacher rather than as a controlling agent.

Old habits die hard. This change in focus may feel awkward for a time, but the payoff will be great.

Summary

In this application of the PCT and HPCT models, I have illustrated the basic concept. I have shown a questioning approach to problem solving which fully respects the other person as an autonomous living control system; facilitating the development of trust, cooperation and high productivity.

> ..you help your associates focus their attention on [work] issues that are important to them... you become their trusted friend, someone who cares.

References

Rijt-Plooij H.H.C van de, & Plooij, Frans X. First published in 1992 as *Oei, ik groei!* by Zomer & Keuning Boeken B.V., Ede and Antwerp English title is *The Wonder Weeks.* Kiddy World Promotions B.V. 2010. For more and editions in 12 languages, see www.thewonderweeks.com.

Runkel, Philip J., *Casting Nets and Testing Specimens: Two Grand Methods of Psychology* (1990, 2007) Menlo Park: Living Control Systems Publishing..

Zuker, Elaina, *Mastering Assertiveness Skills; Power and Positive Influence at Work.* New York: Amacom (1983).

$\mathcal{P}$ERCEPTUAL CONTROL —

LEADING UNCONTROLLABLE PEOPLE

Number three in a series of three articles on PCT.
An early version of this article appeared in
Engineering Management Journal Vol. 7 No.1 March 1995

Perceptual control —
leading uncontrollable people

ABSTRACT

The leadership challenge is to guide, coordinate, direct and yes, control outcomes from combined efforts of associates while respectfully allowing them to direct and control their own experiences. This final article in a series focuses on the hierarchical nature of human experience and extends the application of HPCT to several leadership issues. The testable principles of HPCT have enabled the operation of our nervous system to be demonstrated in working models. The explanation HPCT offers for human behavior can lay a foundation for success of modern engineering management.

Introduction—Leadership and control.

Most corporate leaders and managers are strong controllers. Indeed, leadership and control go hand in hand. The essence of perceptual control is to act in such a way that the result (as you perceive it) of your actions agrees with your intended result. Leadership extends this concept to include actions through others.

The essence of leadership is to control the organization so that the outcome of collective action agrees with the intended outcome. The intended outcome is formulated as a set of specifications in the mind of the leader and goes by many names: want, plan, vision, goal, intention, aim, mission, purpose, target, wish, expectation, requirement, objective, planned outcome. When the leader's perception of the actual results of the organization's collective action agrees with the leader's own specifications, the leader is "in control" and satisfied.

Human control and conflict

We are all controllers. It would be a mistake to think that leaders control and the rest of us (followers) do not control. We all control all the time, making our wants come true. Dwight D. Eisenhower said:

Leadership is the art of getting someone else to do something you want done because he wants to do it.

All living organisms act on their environment in order to experience the environment the way they want it and to keep it that way. It makes no difference if the environment is made up of inert material or living, acting organisms with a "mind of their own"—we do our best to influence our environment to our liking. See *Defining Perceptual Control*, page 15.

The fact that organisms control perceptions and not actions explains why organisms do not need to understand their environment to control it and why faulty explanations are simply ignored in practice. All an organism needs to do is to perceive some variable it controls while it acts on it, and remember which way actions influence the variable. People may say that they act based on some understanding or belief, but if actions suggested by that belief do not produce desired results, people will automatically switch to some other action that appears to work for them. They may not notice that they have switched action and can believe that their explanation is correct. People may be convinced they do one thing while automatically they do another.

Controlling other controllers. Our efforts to influence other organisms—people in particular—are often met with resistance. This resistance establishes conflict. Conflict is a natural result of control by two people, where both try to control the same thing but with different specifications or wants. Conflict also arises when two people control the same thing with the same want but with different perceptions of the actual state of affairs.

It is obvious that it is possible to control inert matter; we do it all the time. But it is not possible to control other controllers without overwhelming force. Without overwhelming force, each organism always autonomously controls its own perceptions. We generally do not employ overwhelming force when dealing with our associates in business. *It is clear that it is not possible to control associates.*

Resolving conflict—the "easy" way. When two controllers with incompatible wants are in conflict, the stronger person (the leader/manager, employer, stronger spouse, parent, police officer or prison warden) can usually cause the other person to cease overt opposition, at least for a while. It may be easy to issue a credible threat against something of importance to the other party, such as loss of income or privileges of some kind. Many managers and hard-charging leaders have discovered that they get immediate attention of associates and can get fast results by threatening something the associate considers very important. The associate will rearrange programs at home and at work and try to adopt the wants specified by the stronger person in order to avoid the threatened consequences.

While the associate may indeed exert effort in the short run, the result of threats may be resentment and loss of satisfaction on the part of the associate in the longer run. This is costly in terms of loss of personal initiative, care, productivity and rapid turnover of personnel.

We all experience conflict and look for easy ways to resolve it, regardless of whether we are the stronger or weaker party. If I am the weaker party in several conflicts, I may become so discouraged that I pretend that my goals and the resulting conflicts are not important to me. Instead I withdraw from conflict when I see it emerging. Withdrawal may be "easy," but the personal price I pay is heavy. I don't learn to control well and conflicts are not resolved. I experience failure and feelings of resentment and despair.

> The fact that organisms control perceptions and not actions explains why organisms do not need to understand their environment to control it and why faulty explanations are simply ignored in practice.

Resolving conflict—the "hard" way. While it may be easy to create the appearance of conflict resolution to the satisfaction of the party with the most clout, it requires care and insight to actually resolve conflict to the satisfaction of both parties, resulting in sustained personal commitment, initiative, productivity and quality. This will be much easier when we learn:

What control is, how it works and what it looks like.

How control causes conflict, what conflict looks and feels like, and how to resolve conflict with mutual satisfaction if at all possible.

Mapping and influencing wants and perceptions (page 34) introduced a way to resolve conflict with mutual satisfaction. Before we can extend this approach and the insights offered by hierarchical perceptual control theory (HPCT) to leadership issues, the hierarchical nature of perceptual control in humans must be well understood.

Hierarchy of perception and control

Levels of Perception and Control. Exhibit 15 summarizes the proposed levels of human perception and control. At the lower levels, the control systems are represented by three rectangles: input, comparator, and output. At the higher levels, memory is inserted. Perceptual signals may be stored in memory and output signals may provide the address to memory which in turn provides reference signals for lower control systems.

Observations by the Plooijs (Van de Rijt-Plooij & Plooij, 1992, 2010) suggest that we are born able to perceive and control sensations at the second level as defined in HPCT. Additional capabilities to perceive and control ever more complex perceptual variables are developed in predictable stages. Each time a new level of perception and control emerges, the first thing noticeable is a regression with uncertainty and anxiety. This may be due to initial confusion and failure of control as the new level emerges.

Failure to control well, at any age, results in large error (difference) signals. Chronic error signals give rise to what we call stress and HPCT postulates that these chronic error signals are sensed by a "dumb" reorganization system, which makes unsystematic changes in the structure and continues to do so until control is reestablished and the chronic error signals disappear or the individual dies—whichever comes first. Thus the concept of reorganization can explain both the successive development of levels of control in infants—as their brains periodically enlarge—and a search for new ways to control in adults under stress.

As an example of how HPCT is a serious, testable explanation of how each of us experience the world and live our lives, consider the rubber band experiment (pages 29-31). I control a visual relationship (sixth level). I don't give any thought at all to my five levels below it, whose smooth, rock solid control keep me upright and move my hand.

As I sit at my desk I see many objects. My visual control systems in coordination with body movements allow me to touch any object at will. My other senses allow me to smell, hear and touch so well that I get the impression that I experience the world directly. The development of the first six levels of perception and control are constrained by regularities in the physical world, and thus are not free to develop any which way and still be able to control. Therefore, you and I develop great similarities in the ways we see, hear and touch the physical world around us. The existence of the lower levels of motor control is easy to demonstrate (page 22).

The story changes as we move up into the more conceptual higher levels. As I read words on a piece of paper, I take their meaning for granted most of the time. That may be a mistake. Words mean something to me as a result of my experience. The same word or symbol may mean something quite different to you as a result of your experience.

It is not easy to recognize how personal my struggle to develop and make sense of the world was and continues to be. Thus I fail to appreciate just how unique my personal perceptions might be at the higher conceptual levels.

There are many ways to hold a fork and knife. I developed one set of memories that now become my wants. I know just how it should feel when I hold those implements: fork in left hand, knife in the right, just like any properly raised Swede would hold them. This is a result of the many social regularities in my home environment. Certain sounds were consistently related to certain experiences, and so I learned Swedish the way my parents, siblings and peers spoke it.

Levels of perception and control	Proposed structure of hierarchical control			Emergence in infants, weeks	Comments, examples - Adult perspective -
11 Systems concept	?	?	?	70-75	Understanding, belief, the way things are, sense of self, identity.
10 Principle				60-64	Generalizations, criteria, standards, priorities, values.
9 Program				49-53	Choices, logical procedures.
8 Sequence				40-43	Simple or repetitive series of events and elements.
7 Category				32-37	"Class membership." Chair, woman. Symbols--words.
6 Relationship				22-26	Walk "on" floor. Bark, dog. Knot "above" target.
5 Event				14-17	Open door, hug, fall, cranking, bounce, reach, grasp, walk.
4 Transition				11-12	Changes in general. Movements.
3 Configuration				7-9	Patterns, edges, texture, posture.
2 Sensation				birth	What kind and how much: Loud, bright, hot, sour, dry,
1 Intensity					Frequency of neural current originated in nerve ending.
0 Environment					Physical effect on nerve ending. Nerve signal to muscle or organ.

Exhibit 15. Levels of perception and control in humans.

Just now I notice that my mouth feels dry. I want it to feel moist. The difference signal results in my system executing a short series of events. You would call it reaching for the cup on my desk and taking a sip of coffee. Suppose I have to go to the cafeteria to get the cup? Now my control systems execute a longer program, branching to several alternative sequences along the way as I make my way past obstacles, get the coffee, and return.

The events, relationships, categories, sequences and programs I observed, experienced and learned from others become the basis for a very personal collection of principles. All make sense to me (well, maybe not, but as long as I don't look too closely, I will never notice and it will not bother me). It matters not at all if these principles stand up to scrutiny of whatever kind. What matters is that I learned them and decided to believe in them. I myself have woven them into systems concepts. All these perceptions stored in distributed memory are mine.

It seems to me that the most important part of my concept of Self is my understanding of the world—the sum total of my principles and systems concepts—my identities.

I hold my own understanding dear. It is me! If someone questions (disturbs) my principles and systems concepts I automatically do what I can to counter the disturbance. For instance, like most people I *hate* performance reviews. A judgement about me by someone else can disturb my concept of self. (An excessive compliment likewise). I resist this disturbance as best I can. If I depend for my livelihood on the judge I may not do anything other than release adrenaline and suffer stress. Perhaps I find a way to dismiss it.

As I have learned HPCT and become more observant of control in daily life, I am impressed with how solid the layers of control are, and how firmly we each control our lives so that we experience the perceptions we want to experience—all the way to the systems concept level. The common denominator is control, not making sense, scientific validity, logic, or any of those niceties. I now see people as autonomous control systems living in a personal world, just like I do. I tolerate personal idiosyncrasies much more easily than I used to.

Going up a level. A good way to resolve conflict, whether it is internal to one person or between people, is to "Go up a level." By this I mean to look at the goals in conflict "from above" and ask what higher purpose is being served by each conflicted purpose at the present level. If the conflict is at the program level of choices, ask about the principle level of priorities. If the conflict is at the principle level of values, ask about the systems concept level of understanding. It does not matter what the level is, how many levels there are, what you call them, if our present labels for the levels are right or if the conflict is internal. Just go up a level; think through the reasons for each conflicted want.

Man's search for meaning by Victor Frankl (1963) featured a good example of this. A widower was distressed over the death of his wife, wishing he had died first. Dr. Frankl asked the man how his wife would have coped if he had died first. By considering his conflicted want from a higher perspective, the widower in a matter of seconds changed his lower level want and decided that he would rather be the one to suffer as survivor. With the want changed, the internal conflict evaporated.

The significance of the hierarchical structure. William T. (Bill) Powers, creator of PCT and HPCT, made these comments to me:

> The first thing a manager has to recognize is that people (including the manager) have many identities, each with its own set of principles and lower goals. A person says "I am a team player," but he also says "I am a competitor" and "I am a Catholic" and "I am a wife (or husband)" and "I am a father (or mother)" and "I am a Republican" and "I am a Rams fan" and "I am a Harvard graduate." These identities have, for each person, meanings that rest on and control the detailed experiences of life at all the lower levels, right down to the color of a shirt and which programs the person watches or doesn't watch on TV.
>
> It's all very well to speak about resolving conflicts in a company, but some conflicts can only be avoided. If one manager is a Catholic and another a Jew or a Muslim, there is no way to align their identities. They pursue different goals at the system concept level. The Rams fan and the Bears fan are not going to compromise and become Jets fans or switch to tennis. The heavy competitor is not going to turn into a team player. All these people may say "I'm a good company man" but they are many things beside that.

Furthermore, there are inherent conflicts in business decisions. As a businessman I realize that since I can't get productivity out of a young and inept worker, I must fire him. As a father, I hate to discourage him. As a Catholic, I want his sins to be forgiven. As a competitor, I rejoice in showing how much power I have over him. As a team player, I want to find a place in the organization where he might do better. In fact, my various identities are in conflict over this person. One of the problems with being in a position of power is that decisions made in one role are abhorrent in another role. Each individual manager has to find a way to support all these different personae; many, unfortunately, do so by compartmenting their lives, which suppresses the conflicts but does not resolve them. A good part of "company" problems are basically personal problems that can't be resolved in a committee meeting. Each person must resolve them alone, particularly at the highest inner level.

All these different identities at the system concept level therefore imply a much larger set of principles, and it is at the principle level where conflicts among identities first become evident. Most business problems boil down to personal problems. One reason that managers set up rigid policies is that by doing so they can avoid the personal conflicts that arise in difficult situations; they can say "I don't like doing this, but it's company policy and I can't go against that." A company (or a family) that is run by rigid policies and rules is an organization in which the individuals have refused to face up to their own inner conflicts or have been unable to resolve them.

Business problems are both technical and personal. Action plans are technical; what is the most profitable mix of products...? Given such problems, willing and intelligent people can come up with answers, try them, improve them, and eventually reach the best solution.

But technical problems interact with personal problems. Technically, it may be best for productivity to shift a product from one department's purview to another's. But this can result in personal problems with people who pursue other goals, such as increasing their influence within the company and ultimately being promoted to a higher level of responsibility—and power. And the principles under which such people work can result in vetoing the technically best procedures, forcing those who work at devising and implementing action plans to look for the second or third best solution, which can result in conflicts elsewhere.

Bill Powers concludes:

HPCT can offer two kinds of useful insights to managers. First there is simply the technical matter of how behavior works; people control for the consequences of their actions, not for the actions themselves. Understanding this can show how to resolve problems that result from mistaking side-effects of another person's actions for intended effects.

But the most important insights come from considering not just that people control, and control consequences rather than acts, but from realizing that the greatest difficulties of organizational life arise at the highest levels of individual human organization. These are the levels where people choose and apply principles as a way of supporting various aspects of their own personal identities. Simply understanding the relationships among logical procedures, principles, and system concepts can help to identify where personal difficulties are arising, which is most of the battle in correcting them.

Applying HPCT to Specific Leadership Problems

Vision and Mission statements. The point of a vision and mission statement is to communicate the organization's aim or purpose. It is intended to help align the efforts of many people toward a common end.

By structuring a statement in harmony with the structure of the human mind, you gain the ability to have every employee understand the basic premises of the organization. Reasons for the important values become apparent, will be questioned, and can be changed as conditions change. Each associate is better able to identify his or her role in the whole and can take initiative to improve overall effectiveness and long term success.

Exhibit 16 illustrates my proposal for the structure of a vision and mission statement as it logically follows from an understanding of HPCT.

Understanding, Belief, Identity

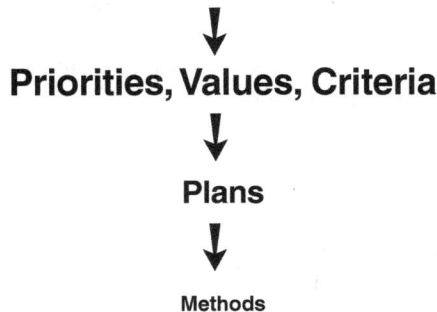

↓

Priorities, Values, Criteria

↓

Plans

↓

Methods

Exhibit 16.
A hierarchical vision and mission statement.

Understanding, Belief, Identity can be stated as "These are the facts as we understand them and this is our identity." This statement describes the organization and the world in which it operates. Problems, opportunities, unique capabilities, personal convictions, resources, technologies, etc. This should not be a nebulous, feel-good, statement of good intentions but rather extensive, multi-faceted documents every associate can study, learn from, and take issue with as conditions change and the factual descriptions need to be updated. It takes time to internalize principles and systems concepts from experience and update them from current reports on markets and technology. Some companies have commissioned books to tell the company's history in order to provide new associates with background information.

Priorities, Values, Criteria follow: "Therefore, we conclude that this is important." This is a statement about prioritizing choices, values, and standards that can be agreed to by leaders and associates alike.

Plans spell out: "These are the results we want, this is what we want to see." Plans are developed for each functional unit within the organization, based on priorities. Since perceptions, not action are what living organisms control, plans should focus on results under various circumstances, not micromanage action.

Methods amount to planning in more detail: "These are the interim results we want to see on the way to final results."

With these statements, prospective associates can align their own personal understanding and priorities with the legitimate needs and clear priorities spelled out for the organization and their local group. Good information at the principle and systems concept levels becomes a powerful guiding force. A clear context is created for the organization that gives meaning to job descriptions at the action plan and methods levels. Note that the word vision is essentially synonymous with goal, aim, and want. Thus vision applies at several levels.

Performance coaching reviews. Traditional performance reviews are institutionalized with the best of intentions. Reviews are designed to provide feedback and help associates improve and develop. But everyone hates performance reviews. Why?

When an associate is presented with a judgement, this can be a forceful disturbance to some aspect of the concept of self. Such a disturbance cannot be countered effectively (without risking employment). A judgement does little to enhance a person's capability to perceive or choose wants effectively, to be more effective and capable of satisfying himself or herself or the organization.

An alternative to the traditional review is *performance coaching*, a procedure that follows naturally as a variation of *mapping and influencing wants and perceptions* (page 34). Respect requires that the associate be in control of the performance coaching as much as possible. This review should be conducted at least once a month.

1) Schedule a regular, undisturbed meeting.
2) The associate begins the review by submitting a handwritten or typed description of one or two projects, challenges, or situations he or she has dealt with since the last meeting. Details can be embellished orally. This leads to a supportive and appreciative discussion, focusing on current job issues of whatever kind.
3) The associate describes one or two areas where he or she believes improvement is needed. The manager can raise some issue, too.
4) The manager works with the associate to formulate a plan for improvement that the associate can carry out with support as needed.
5) The associate and manager both commit to follow up on the plan.

Helping in this way is far more proactive and supportive than to wait and see for 6 – 12 months and then judge.

1) The associate learns from the manager.
2) The associate performs better.
3) Mutual satisfaction and trust develops.
4) Managers get a thorough and realistic picture of what the associate is doing, is capable of doing, and where assistance is needed.

When discussions about performance are perceived as normal, non-threatening, commonplace events, people will relax. Both parties can talk freely about expectations, goals, and problems. The focus is on assisting one another to be more effective and satisfied. This is interaction with mutual respect. Trust develops naturally. When the time comes to consider a promotion, management has a detailed, personal record of the associate's capabilities and progress.

Develop team spirit and caring relationships. Consistent application of the principles of HPCT to conflict resolution, vision and mission statements, and performance coaching will go a long way towards creating trust and a sense of belonging in an organization. Productive teamwork requires careful integration of all these leadership applications. Jim Soldani has reported on his early application of HPCT to leadership of a team of 120 people in a manufacturing environment (Soldani, 1989). In seven months, the measure of work completed on schedule went up from 23% to 98%. Overtime declined from 12% to 3%. Quality went up by a factor of 5. Work in process inventory fell by a third. Productivity went up 21%. Total savings added up to about 1.5 million dollars a year.

Development and coordination of goals takes time, and resolving conflicts requires personal involvement. Jim's application of HPCT paid off. His team won awards repeatedly for excellence with high productivity.

Individuals in a team share a common goal, which becomes the team's focal point. Many different kinds of goals can qualify as a focal point goal, for example, "customer satisfaction with our services," "performance to schedule," "production cycle time," and "quality of which we can be proud." Such goals involve perceptions that can be qualitatively defined, quantified at some level, measured on an ongoing basis, and thus possibly be controlled.

While each such goal may be reduced to one measure, it naturally breaks down into subgoals. You cannot perform to schedule if parts are not available or don't fit. You can't be proud of your quality unless your customer is satisfied and you have incorporated your best know-how into your service even if your customer did not specify it.

The focus goal must be broken down with great care into subgoals each team member can perceive and control with the help of appropriate information and resources. To join the team, each member is asked to adopt the focus goal and appropriate subgoals.

As work progresses, conflicts of all kinds may surface. Some may arise from the work, i. e. conflicting methods, resource allocations or subgoals. Some may be personal, where responsibility to the team conflicts with personal preferences, work habits and emergencies: "Our shipment is late and I could make up for it if I stay late, but tonight is my bowling night." "Joe didn't do his task, so that's why I didn't get my part done." "Joe didn't do his task, and when I realized it, I worked two hours overtime to make up for it." "I need to take my wife to the hospital tomorrow. Can you cover for me?" Team leaders can help associates sort out conflicts, clarify responsibilities and make accommodations by *mapping and influencing wants and perceptions*. Team leaders can also help junior team-members learn to control and perform well through *performance coaching*.

Most people crave the company of others. Consider Exhibit 15. The first six levels of perception defined here can be called "experience levels." As we live, move about and work, we experience the world around us with a rich assortment of senses. (Communication by words alone starts at level seven and is nowhere near as rich in sensory detail). As I work in the vicinity of people I like, I experience my associates and develop a positive perception of them individually as worthwhile human beings. If I don't like some teammembers, I can still respect our mutual commitment to the focus goal and we can both take professional pride in achieving and maintaining our subgoals. This builds a sense of belonging, human connection and team spirit. Relationships are strengthened by awareness of each other as we cooperate in productive work.

For a theoretical discussion of teamwork based on an understanding of the HPCT model, see *CT Psychology and Social Organizations* (Powers, 1992).

Total Quality Management. TQM comes in many flavors, all of which can be linked to HPCT. Consider the four areas of Profound Knowledge defined by Dr. Deming:

1) Appreciation for a system
2) Knowledge about variation
3) Theory of knowledge
4) Psychology

HPCT supports TQM by offering a theory of psychology that can be tested and clarifies experience.

Going beyond psychology, I observe that the family of statistical tools that are usually thought of as the foundation of TQM (SPC charts, fishbone diagrams, flow charts etc.) all serve to improve our perception of controlled variables in production processes. You cannot control what you cannot perceive. When you give the worker measuring tools and SPC charts, with clear specifications of expected outcomes, you immediately improve the workers ability to control production. Results have been impressive.

While many associate TQM with statistical tools, my conclusion is that the essence of TQM consists of well chosen aims and good perception in a fully functioning (control) system.

In 1993 I had an opportunity to introduce PCT to a Deming user group. The resulting two hour video (Forssell, 1993) features a discussion of Jim Soldani's results from applying PCT in manufacturing, a live demonstration of the rubber band experiment, and my interpretation of the essence of TQM and Deming's fourteen points.

Non-manipulative selling. Some may say that sales has nothing to do with leadership. Others may say we sell at all times, and that sales is the essence of leadership. Regardless of whether we deal with people inside or outside our own organization, in a regular working relationship, in a service capacity or in outside sales, the question is this: "What can you do when you want to influence and lead someone you don't know yet, someone who does not know you, or a superior or associate you find hard to approach?" Let's call this someone a prospective customer.

To approach an associate with the process of *mapping and influencing wants and perceptions,* you might simply say: "Can you come into my office? I'd like to talk to you. Is this a good time for you?"

The difficulty that requires Non-manipulative selling is that (unlike a close associate) your prospective customer may have no reason to hear you initially. Your prospect is focused on other things and may not be aware that the capability you offer exists. Therefore, as a salesperson you must make a careful approach. Please review Exhibit 13. Person 1 on the left is your prospect. Person 2 (and 3 and 4 etc.) on the right is the prospect's associate in their organization.

Your prospect, person 1, functions in an environment, interacting with the other people in the organization. Why should your prospect read your advertisement, read your letter, talk to you on the phone or talk to you in person?

Well, looking at Exhibit 13, what variables are your prospect controlling? Are they being disturbed? What are the likely wants? What are the likely perceptions of the variable? How do you think the comparison looks to your prospect—what are likely error signals? In other words: What is your prospect concerned about? Consider your prospect's position and industry in relation to your offering and develop a script with questions exploring concerns your prospect might have.

A salesperson is a teacher. Here, you teach person 1 to control better, with greater productivity and satisfaction, incidentally using your idea, service or product.

The process becomes:

1) Guess what the prospect may be concerned about.
2) Gain attention and interest by relating to those concerns, perhaps by telling a short story of how you have helped someone else control better in similar circumstances. Offer more information.
4) Follow the methodology of *mapping and influencing wants and perceptions,* first with your prospect and then with other associates in the organization as required until you have shown them all how to control their lives better with the help of your offering. You will not have to ask for the order; your prospect will ask to buy.

A technical summary of lessons for leaders

I have pointed out that it is not necessary to understand anything, including control, to live and control, since we *are* controllers. But if you *do* understand how and what people control, you can be more effective as a leader. From the detailed insight of HPCT, leaders and managers can learn several lessons:

1. Leaders and followers alike act only to produce and maintain intended perceptions. How people act in order to do this is determined in part by their environment. People control *results* (their perceptions of outcomes), not the *means* used to produce those results (their actions). People achieve consistent ends by variable means. Because associates control what they perceive, the first task of a leader is to ensure that everyone is able to perceive the common goal in terms of all the perceptual variables that make it up; what the multiple dimensions of the goal are, and

which dimension each associate should control to avoid conflict with team mates controlling other dimensions. In the language of TQM, we talk about the importance of shared operational definitions. What *kind* of goal and subgoal?

2. Once people know what perceptions to control, they must know the appropriate levels at which to control them: Leaders must establish common reference levels for the intended states of the controlled perceptions—how much is desired in each dimension? This means a clear specification of the desired result; a target or goal that can be achieved continuously as part of a process or as a step in a chain of events. How *much* of each kind?

3. All associates are controlling a whole constellation of perceptions; the perceptions to be controlled at work are just a subset of the perceptions people control. Leaders are wise to be sensitive to the fact that control of certain variables will conflict with control of other variables. Be *flexible* about who does what, when and how.

You now understand that associates control perceptions, *not* actions. You are there to help people understand *what* results are to be produced—not *how* they are to be produced. HPCT shows why "micro-management" creates conflict, is resented, and produces poor results.

4. When you understand how wants relate to understanding and priorities, you can design a corporate vision and mission statement that comes alive with meaning and from which people can easily derive their own personal mission statements.

5. When you understand the source of associates' emotions, you understand how they can change.

6. When you understand the role of personal "worlds" of perception and memory, you can anticipate conflict and create cooperation by *mapping and influencing wants and perceptions.*

When you teach HPCT to your associates, everyone can use the same understanding and approach, dealing with people at all levels, inside and outside the organization.

Summary

The leadership challenge is to guide, coordinate, direct and yes, control outcomes from combined efforts of many associates while respectfully allowing them to direct and control their own experiences freely. HPCT offers a new explanation for human experience. It is technically elegant, conceptually simple, testable, and better than "common sense." HPCT is really an "engineering theory," but its principles are readily understood by any attentive person. You can apply the explanations of HPCT to past experience as well as thinking ahead. Your own experiences suddenly make more sense to you, and you get a new perspective and capability to support associates, employees, vendors, customers, friends and family in individual, autonomous exercise of well-informed, effective and satisfying control.

In this series of three articles I have introduced the basic principles of HPCT and outlined how they can be applied to issues of leadership and management. HPCT provides a deep understanding of human nature, and of the problems in getting groups of individuals to work in concert for their own satisfaction as well as that of their organization and customers. When leaders and followers alike understand the basics of HPCT, they all understand that each person is inherently purposeful, and that each is responsible for maintaining his or her own integrity in an organization. Agreement on goals and elimination of conflict gets easier.

Every application of HPCT to a specific company with its specific problems has to be worked out on the basis of this deep understanding; there are no formulas to apply, but only principles of analysis that will lead to specific solutions. As more managers understand and try the principles of HPCT, we will all learn more about how to apply them. As the number of people exploring the uses of HPCT grows, our understanding of the productive community will grow.

Acknowledgments

In these articles I have drawn on the experiences of several good friends for inspiration and validation. I consider myself fortunate to have come across the writings of Bill Powers. Bill is a warm human being, an untiring champion of clear thinking, and a patient teacher. Ed Ford is a wise counselor who has shown how HPCT gives insight into why his approach to relationship counseling and school discipline work so well. Jim Soldani combined his experiences as a manufacturing executive with what he learned from Bill Powers and Ed Ford with very good results in every dimension.

I met Mike Bosworth ten years ago, well before I found PCT and learned Solution Selling® from him (Bosworth, 1995). While solution selling has developed from traditional research and experience without knowledge of HPCT, I am satisfied that the reason salesmen trained by Mike are successful is that Solution Selling teaches them a way to focus on the prospect's wants and perceptions much the way I have outlined here.

References

References listed previously are not repeated here.

Bosworth, Michael T., *Solution Selling; Creating Buyers in Difficult Selling Markets.* NY, NY: McGraw-Hill (1995).

Forssell, Dag C. *PCT supports TQM.* DVD, 2 hr (1993). Menlo Park: Living Control Systems Publishing.

Frankl, Victor E., *Mans Search for Meaning*, New York: Whashington Square Press (1963).

Powers, William T., *Living Control Systems, Volume II: Selected Papers.* New Canaan, CT: Benchmark Publications Inc. (1992).

PERCEPTUAL CONTROL — DETAILS AND COMMENTS

This paper is the original draft for the first half of the second article:
Perceptual Control — management insight for problem solving.
The article was rewritten to to focus it on problem solving.

Perceptual control — details and comments

INTRODUCTION

These comments provide more background and perspective on how PCT is different from contemporary psychologies, and develops the architecture of HPCT suggested in the first article, exhibit 4.

How is PCT different?

Let us contrast PCT with the linear cause-effect perspective of contemporary schools of psychology.

First, let me ask you: What is the most common explanation for why people behave? People respond to stimuli in their environment, right? How they respond depends on how they have been shaped or conditioned by their environment. This means that what happens to people determines what they do.

Some management programs tell you how to push people's "buttons" so they do what you want them to do. Some programs advise you to assess what situation you are in to know which behavior to use.

Some sales training gives you a choice of "17 different ways to close," depending on how you read the customer's situation and attitude. Of course, you must know what situation you are in and what buttons to push.

Would you agree this doesnt work all the time?

Another explanation is that our thoughts, our plans and decisions determine what we do. As an example, think of how you play solitaire. You sit quietly and think. There is no stimulus from outside you. You decide to place an ace on a king and do it. This is a cause-effect perspective too, only with an internal cause. This, too, appears true some of the time, but does not work all the time, because it is not the whole story either.

> The basic postulate of PCT is this:
> ## *it's all perception*
> We experience the brain's perceptual activities, not the world itself.

PCT gives you a complete picture of how both the environment and internal goals relate to action. PCT provides diagnostic tools that help you see how a system of perceptions, goals and actions is working continuously. This means that you can always understand the structure of functional interactions in yourself and in others, and can figure out what questions to ask to learn details at any time.

If you do something that works well, PCT explains **why** it works. If you are doing something that does not work well, PCT will indicate why and suggest new approaches. For example, if you use a wise, principle-based management program, PCT will make it more understandable and easier to teach. If you use a respectful, non-manipulative sales approach, PCT will make that more understandable too.

There are many natural leaders, successful salesmen, wise parents and good communicators. But they cannot explain what they do in any depth. Their insights and skills are intuitive. PCT provides the missing explanation.

> Without an understanding of control, [people] are literally blind to a phenomenon that is fundamental to all living organisms.

Hierarchical Perceptual Control Theory

Exhibit 17 shows more of the architecture first presented in exhibit 4. The two dimensions of this model of the human mind are:

1) Levels of perception and control and
2) Examples of perception.

You will find that thinking in terms of these two dimensions is very helpful when counseling associates and resolving conflict. A control system at the center of exhibit 17 (❖) has been highlighted. The demonstration that follows shows how you can focus your attention on control of something, in this case a single visual relationship, and how your mind makes it come true, working through your entire body. But first, let us examine how the proposed architecture works.

Levels of perception

The *vertical dimension* in exhibit 17 is "Levels of perception." Exhibit 18 shows more detail. Starting from the bottom, a low-level input—a neural current created by a nerve ending "tickled" by some physical phenomenon in the world, such as light falling on a single cell in the retina—is combined with other inputs, creating a perception signal at a higher level, which is in turn combined to create a signal at a still higher level. At the higher levels, a branch of the perceptual signal can be recorded in memory and later played back as a reference signal. (It is beyond the scope of this article to suggest an integration of distributed memory in HPCT, with suggested explanations for imagination, automatic control and passive observation).

Levels of perception are central to HPCT. They were introduced by Powers (1973), and have been further described in detail by Robertson and Powers (1990). Some of the computer demonstrations show how how hierarchical control of perception works, and the file PerceptLevels.pdf posted at PCTresources.com explains how to think about the levels. I will not describe the proposed levels in detail in this introduction, but the basic postulate of PCT, simply put, is this: *it's all perception*. We experience the brain's perceptual activities, not the objective world itself.

Levels of perceptual control

Exhibit 19 incorporates exhibit 18, and completes the picture with control at the same levels as perception. This arrangement is shown in exhibit 17 in the two areas of muscle action and physiology, but not in the other senses. All the control systems shown in exhibit 17 act on the body outside the brain through both muscles and physiology, and on the world outside the body through muscles.

You can think of the chain of control systems in exhibit 19 as an organization with a worker at the lowest level, a supervisor at the second level and a manager at the third level. An equivalent metaphor is to think of a driver and two rows of backseat drivers. The driver (control level 1) sees the road through a TV screen and does the steering. The driver at level 2 gets a summary report passed on from an interpretation of the driver's TV, combined with summary reports from other TV's. The driver at level 3 has similar options. You can easily imagine that this third level driver combines wants of his own superiors of different "Examples of perception," then shows where to go by selecting a map in memory. The second driver reads the map and specifies which streets to use. The first driver converts these more detailed instructions into control of positions through action—turning the wheel. If the communication is fast and reliable enough, this arrangement will work fine in real time.

The human body has about 800 muscles. Therefore, the muscle tension control chain in exhibit 17 represents at least 800 interconnected control units at the first level. When you walk, you may address a memory stored at the event level, which holds a certain walking pattern. This memory plays back a reference signal which is converted with additional inputs at the transition level into a certain speed of this walk. The configuration level converts this reference signal into smoothly varying leg positions, which result in changing velocities at the sensation level. Changing velocities require changing accelerations, which the tendon reflex loop*, at the intensity level, accomplishes by varying muscle contraction. If your toe hits an obstacle, the limb acceleration, velocity and position are disturbed. Within fractions of a second, the tendon reflex loop compensates by changing the muscle force. This explains why you recover from a

---

* See exhibit 25 in *Are All Sciences Created Equal?*

stumble even before perceptions of the stumble have been combined and reported all the way up your internal "chain of command"—as shown in exhibits 2 and 3. You don't just react with some mysterious reflex—all your body's muscles are under exquisite control at all times. When you specify a perception at a high level the hierarchy delivers a real time perception very close to what has been specified by acting on your environment. The HPCT term for this is that perceptions behave. *The Hierarchical Behavior of Perception*, (Marken, 1993) reports on this in greater detail including response times in humans.

Examples of perception

The *horizontal dimension* in exhibit 17 is "Examples of perception." At the lowest levels, we perceive light, vibration, pressure, temperature, joint angles, tendon stretch, smell, taste and physiology (which we sense as a part of feelings). The highest perceptual levels are called systems concepts. These are descriptions, explanations and models of the world, in many areas of knowledge, which we learn and decide to believe in, as exemplified in exhibit 17. Patterns of principles and systems concepts taken together constitute what we call culture, science, religion, ethnicity and so forth.

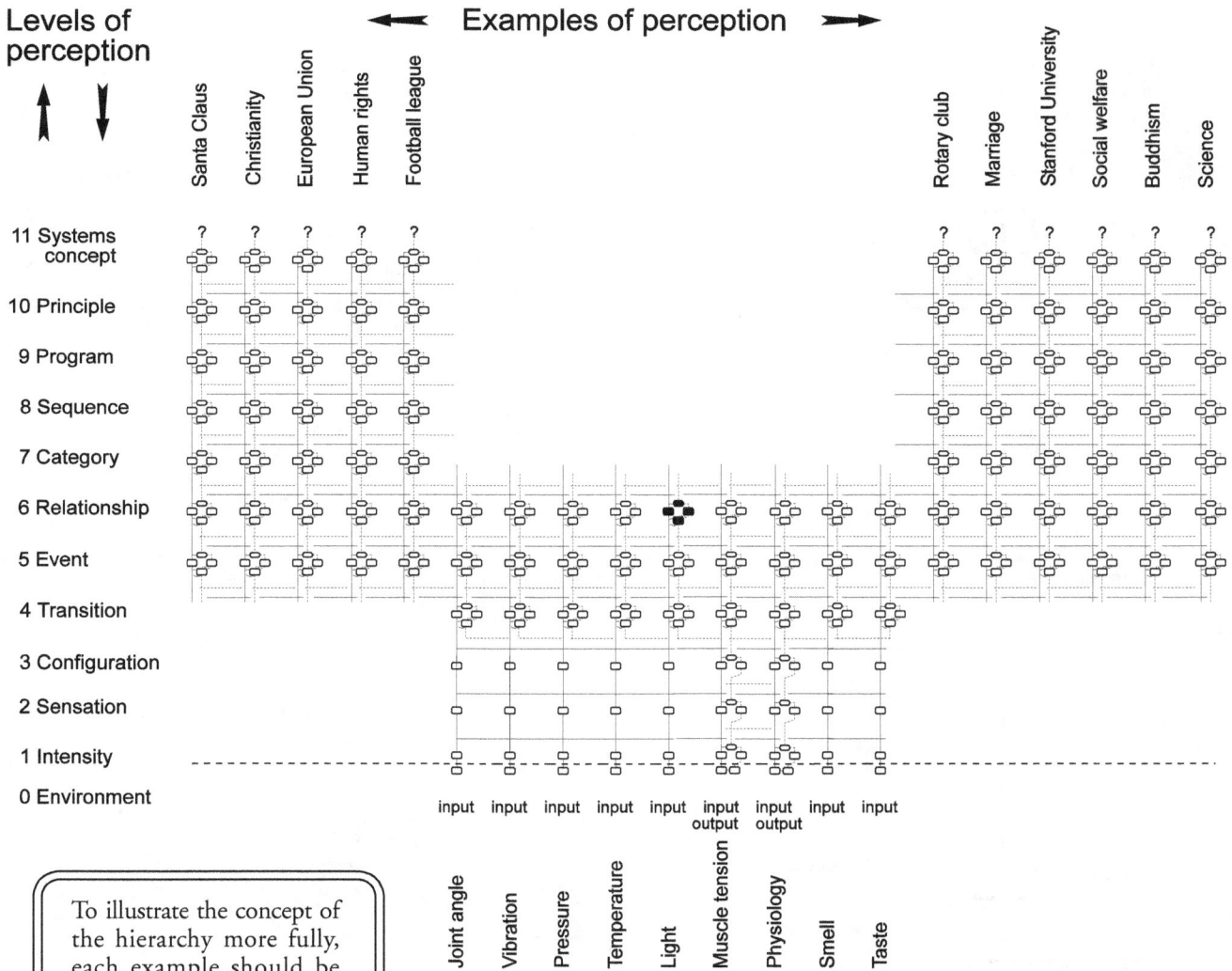

Levels of perception ← **Examples of perception** →

Levels of perception		
11 Systems concept		
10 Principle		
9 Program		
8 Sequence		
7 Category		
6 Relationship		
5 Event		
4 Transition		
3 Configuration		
2 Sensation		
1 Intensity		
0 Environment		

To illustrate the concept of the hierarchy more fully, each example should be labeled at each level.
For an early suggestion of such labeling, see Living Control Systems I, p. 206.

Exhibit 17. Conceptual illustration:

A person as a hierarchy of interacting control systems. (Inspired by an illustration created by Mary Powers.)

The insight HPCT offers is that these principles and systems concepts are perceptions in themselves. In daily language we talk about understanding, belief, or generally "the way the world is or should be."

Based on the *systems concepts* we have internalized, in comparison with the world as we see it, we select *principles* to live by: priorities, values, standards. These in turn, again in comparison with perceptions of the current world, determine the *programs* or action plans we carry out. From these follow *sequences,* or methods made up of *events,* work elements needed to carry out the programs we have chosen. Events require control of muscles and body chemistry at the lowest levels.

With this brief outline, I hope you can see how your own perceptions "behave" all the way from your highest systems concepts down to muscle fibers and chemistry. You don't have to have a detailed outline of HPCT to realize that what you really want—what is important to you—you make come true as best you can. We control our world as we perceive it from the time life began until we die.

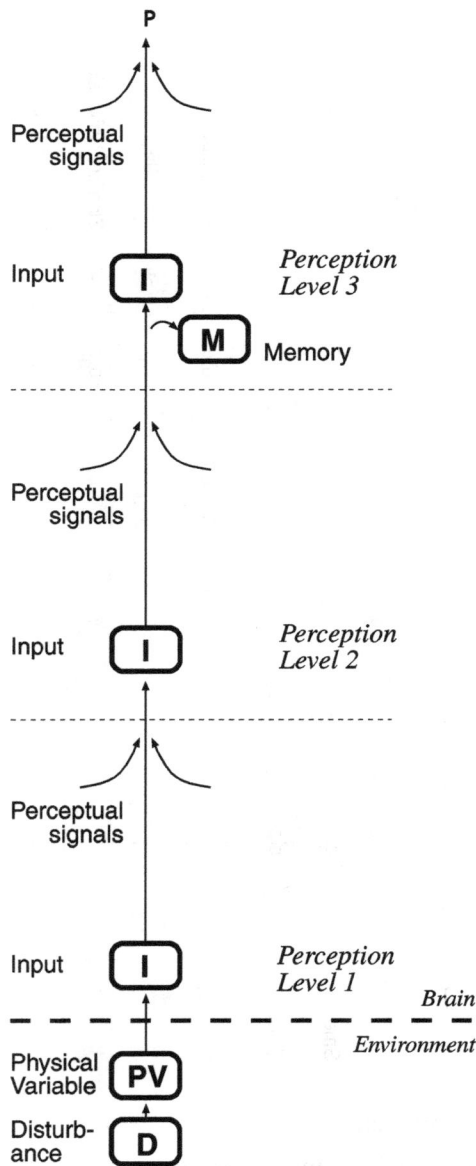

Exhibit 18. Levels of perception

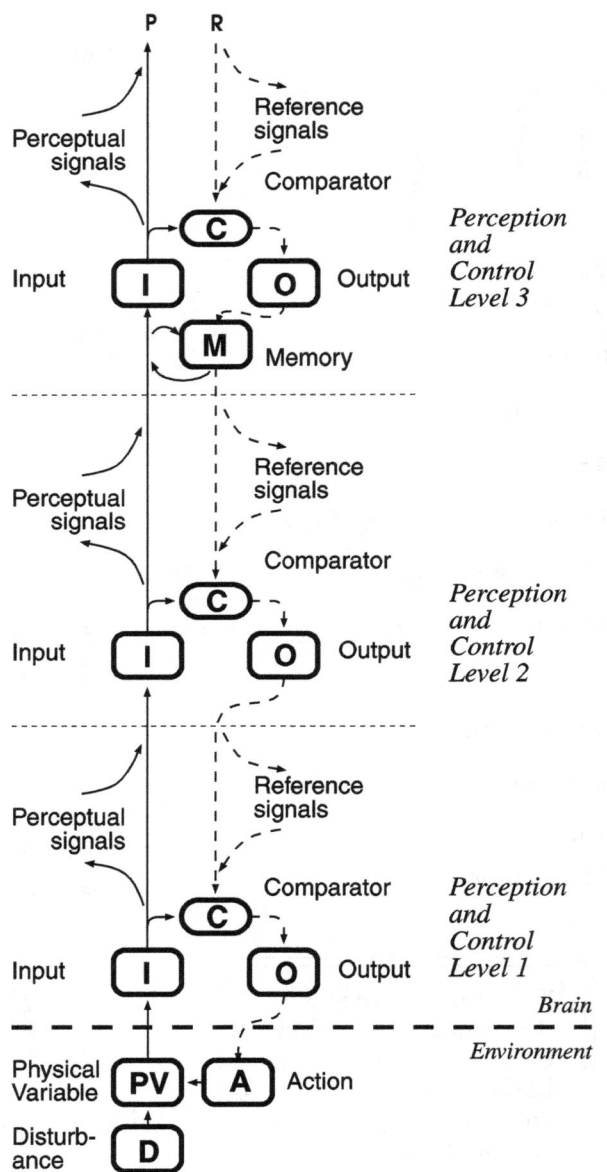

Exhibit 19. Levels of perception and control.

> When you specify a perception at a high level the hierarchy delivers a real time perception very close to what has been specified by acting on your environment.

When as manager, teacher, parent or friend, you want to help people control their world better, to be more effective and satisfied, exhibit 17 suggests that one of the things you can do is to help people improve and expand on their internal control capability by clarifying and developing their perceptions at higher levels, in relevant subject areas (see *mapping and influencing wants and perceptions*, page 34). The world portrayed in exhibit 17 is internal to a person's mind. A person is the **only** one who can question the validity (from her own point of view, of course) of the perceptions stored in her own mind. Therefore, a good way to assist people is to ask them questions about their systems concepts, principles, and programs in the areas or subjects of knowledge that is relevant to their problem or conflict; questions which help them talk to themselves.

When you respect people as autonomous living control systems, you realize that you cannot impose your opinions. You will not be surprised if they ignore you when you try. You gain trust when people they realize that you are helping them control their lives more effectively. *Freedom From Stress*, (Ford, 1989) is a very readable introduction to PCT that illustrates these principles with a counseling story and roleplays that touch on work, marriage, family, and school.

Early development

An exciting aspect of HPCT is that it provides a rational, consistent explanation for our development all the way from conception to adulthood using the same basic building block of control. An infant has developed some ability to control both muscles and physiology. The fetus has been able to hear, taste, touch, smell and move about, and thus practice these perception and control capabilities, but has not experienced vision, nor coordinated it with eye and body movement.

In their article: *Developmental Transitions as Successive Reorganizations of a Control Hierarchy* (Marken, 1990), Dutch researchers Frans X. Plooij and Hedwig C. van de Rijt-Plooij report their observations of mother-infant development among free-living chimpanzees. They identify and describe progressively higher levels of control capability (giving examples all the way up to the emergence of the principle level at about 18 months of age), with short periods of regression and crisis between them, as if the infant takes one step back and two forward to develop. They note that movements are rapid in the beginning, as when the newborn roots for the nipple on the mothers breast, and slow down as higher levels of control develop and the infant no longer searches by means of sensing temperature of the skin and nipple (second level control—sensation), but instead perceives the visual image of breast and nipple (third level control—configuration), then moves directly to the nipple, but more slowly. This is consistent with the engineering requirement that higher control systems be slower than lower ones. If they were not, the hierarchy could not be stable.

The Plooij's have later studied human infant development (Rijt-Plooij, 1992, 2010), and have identified 10 highly predictable periods of mother-infant crisis in the first 18 months of life. They have found that the newborn infant controls at the second level, with perception of configuration emerging at 7-8 weeks, perception of transitions at 11-12 weeks, events at 17 weeks, and so forth. The principle level emerges at 14.5 months and the systems concept level (including the notion of self) towards 18 months. Their book *Oei, Ik Groei! (Wow, I Grow!)*, based on this work, is written for all parents and reports both on infant development and the mother-infant conflicts that go with it. It is easy to understand, very practical and became the top nonfiction book in The Netherlands in 1993. *The Wonder Weeks: How to stimulate your baby's mental development and help him turn his 10 predictable, great, fussy phases into magical leaps forward.* Available in twelve languages. See www.thewonderweeks.com for details.

> You gain trust when people realize that you are helping them control their lives more effectively.

Reorganization

When an organism (young or old) fails to control its world well, perhaps due to conflict, large differences (error signal, dissatisfaction) arise between what the organism wants and what it experiences. This large error signal creates large neural and biochemical signals**. HPCT postulates that such chronic error signals are undesirable and that they are perceived by a very basic, "dumb" biochemical control system which as its output causes random changes in the organization of the control hierarchy. This is called *reorganization*. It is thought to take place at a basic neurological and biochemical level as well as at the high levels of principles and systems concepts, and explains both the development of infants and changes in adults, even dramatic ones. The idea is that chronic error and reorganization (being random, it can be good or bad) continues until some change happens to rearrange the control systems in a way that works better. At that point, the chronic error and reorganization both stop. The process of reorganization manifests itself as crisis, frustration, and discomfort. Many different neural and biological rearrangements may be tried until something serves to restore control or the person eventually dies. We recognize mild reorganization when we have a complex problem that troubles us. Our mind churns ideas and we say: "Let me sleep on it, a solution will come to me." A manager can support an associate who is reorganizing by explaining the process, reassuring the associate that (most of the time) there is light at the end of the tunnel and, if asked, offer more effective ways to perceive the situation and more effective choices to make, thus reducing the randomness of the process.

** See also *PCT explains feelings* (page 71).

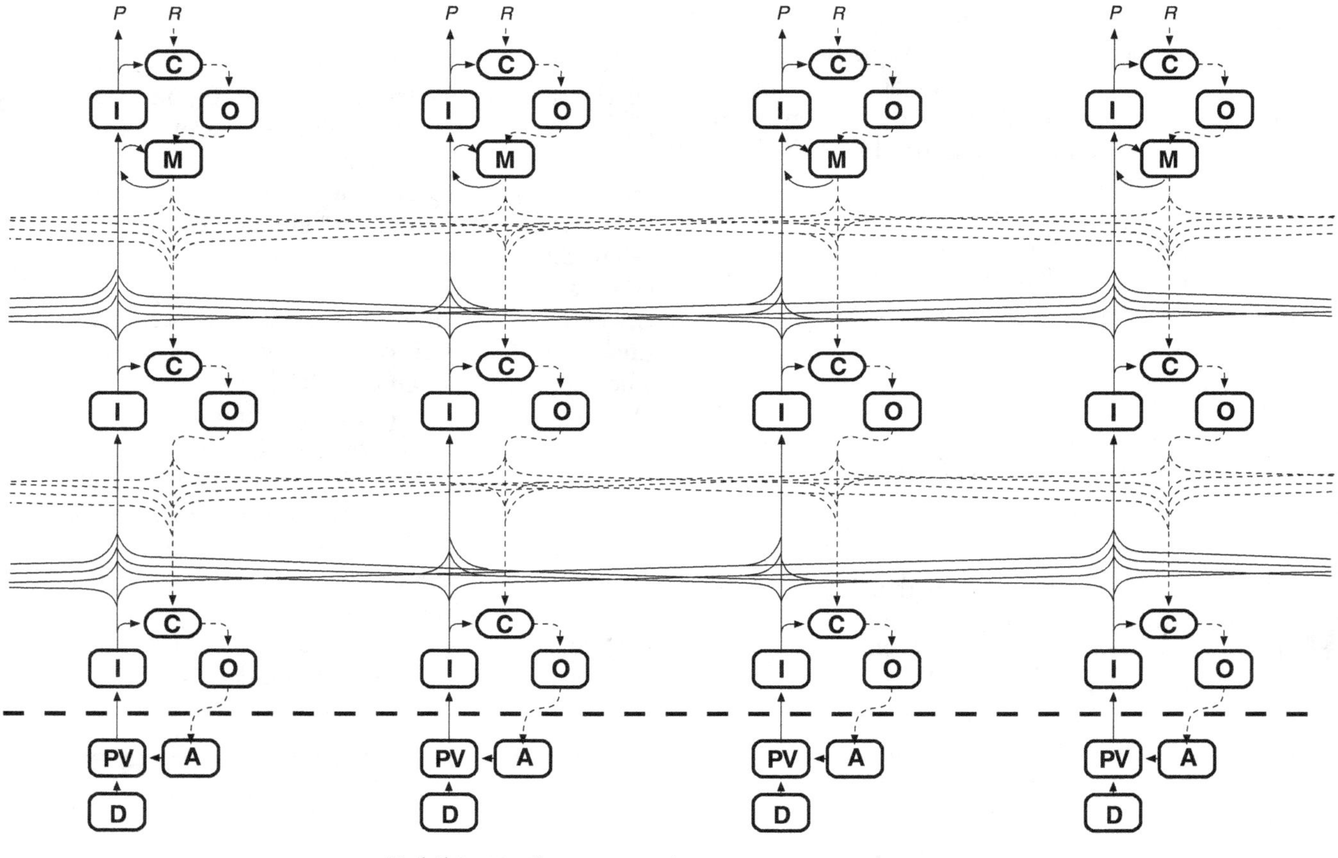

Exhibit 20. Interconnections among control systems.

Interconnections

The horizontal lines shown in exhibit 17 represent connections with other control systems, both adding perceptual signals together as higher-level perceptions are formed, and distributing reference signals to several lower-level control systems. Exhibit 20 suggests the full complexity of such interconnections. Such hierarchies can both be stable and satisfy many different high-level specifications. In this illustration, each of the four low-level "workers" work on a different process to satisfy the combined demands of four different intermediate "supervisors" who in turn satisfy the combined demands of four different "managers." This sounds like an impossible nightmare in terms of a matrix organization in business, but is clearly illustrated by the spreadsheet simulation that can be downloaded at www.mindreadings.com. This simulation shows that the control systems either

a) converge on a stable "worker" solution that satisfies the disparate demands of both "supervisors" and "managers" quickly and efficiently, or

b) develop severe internal conflict with large outputs which cancel one another, maintain chronic errors, and waste energy.

One real world application of this kind of capability in a human being is the maintenance of physical balance. We don't usually think of a human being as a tower made of sticks, swivel joints and active rubber bands carrying out a balancing act all day long, do we? When you stand at a blackboard and write, you focus on your hand movement. But hand movement upsets your balance, so in order to maintain that specification at the same time, most of your skeletal muscles are continuously compensating. You cannot stretch out your hand without the muscle in your big toe getting involved, can you? Exhibit 20 illustrates the Spreadsheet demo which provides an active demonstration of how smoothly a hierarchy of control systems can take care of multiple demands without your giving it any conscious thought at all. While you are still at the blackboard, select a memory that specifies some rhythmic changes in your balance and position, and you find yourself dancing, still maintaining harmony and cooperation among all 800 muscles in your body. A hierarchy of control systems is simple and does the job out of sight and (most of the time) out of mind.

A demonstration of control

You are now turning your attention to the highlighted control system in the center of exhibit 17. Notice how all the control systems in your hierarchy connect your visual experience and difference signals to your muscular control systems, which move your hand while maintaining your balance.

— — — — — — — — — — — — — — —

Here ends the original draft for the first half of the second article. The rest of the article continues on page 29, right column:

A demonstration of control

— — — — — — — — — — — — — — —

For more demonstrations, I suggest *Portable PCT Demonstrations* (Greg Williams, ed) in *Closed Loop*, Spring 1993, Vol 3 No 2.

> We control our world as we perceive it from the time life began until we die.

ARE ALL SCIENCES CREATED EQUAL?

Are all sciences created equal?

Dag Forssell 1994

ABSTRACT:

Sciences of today are not created equal. The physical sciences we depend on today were not always dependable. The life sciences we cannot and should not depend on today may become dependable in the future. While *Perceptual Control Theory (PCT)* deals with a "fuzzy" subject, it differs from contemporary life science in the kind and quality of explanations offered. To clarify this difference, categories of experience and explanation are defined and illustrated. PCT is not explained in this paper, but perspectives, basics and some explanations are discussed.

Introduction

I have long been interested in "what makes people tick." When I read *Behavior: The Control of Perception* by William T. (Bill) Powers, the detailed, in-depth explanations made perfect sense to the engineer in me. Demonstrations were compelling in their universal application and validity. I found the book very different from seminars, books and tape programs I had studied before. Powers provides a lucid synthesis, showing how neurons interacting in a hierarchy of control systems can account for most of the phenomena we experience.

I found that applying my understanding of PCT can help me develop and maintain pleasant, productive personal relationships on and off the job. PCT shows me that I am an autonomous living control system, and I value my ability to control my perceptions freely. In my roles as father, husband, friend, teacher and manager, I now strive to support others, especially those close to me, to control their perceptions in a way that is satisfying to them. This motivates me to teach PCT.

I have become acutely aware that PCT has been distorted, misunderstood, oversimplified and dismissed by scientists who deal with the descriptions and explanations PCT improves upon—or replaces. I have wondered why some people grasp and appreciate PCT with ease while others find it difficult to understand, accept or both. There appear to be two reasons for this.

The first reason, well explained by PCT, is that once a person has been taught an idea and decided to believe in it, that idea becomes part of the person's control hierarchy and any suggestion that the idea is false is resisted. Kuhn (1970), shows how this has been true for many scientific revolutions. Any adult has woven a personal web of ideas of how people "work." Suggestions that don't fit this web of principles and systems concepts are quite naturally resisted—or misinterpreted or distorted so they do fit.

A second reason may be that there are significant differences between the kind of theory and explanation scientists are used to in different fields, and that these differences make comprehension difficult. Scientists who are used to deal with descriptions alone may fail to understand the kind of explanation PCT offers. In this paper I address this second reason by discussing theories and explanations. "Theory" can mean anything from a hunch to a law of nature. I propose the categories *Experience, Description, Descriptive Non-Explanation*, and *Causal Mechanism,* and shall point out the advantages of causal mechanisms.

Language and expectations

We like to say that we live in a scientific age. Every day newspapers and TV programs announce new findings by scientific researchers. Scientific research done by a scientific method suggests definitive information, double-checked by researchers and 100% valid. This interpretation may be overly generous. All sciences are *not* created equal. Some very important fields of science are not very scientific at all, lacking explanations that have proven valid.

Theory and science go together, but in popular usage the word *theory* can mean anything from a casual hunch based on personal experience (which is hard to articulate) to a law of nature which has been confirmed in innumerable rigorous experiments. A *paradigm* means any personal way of looking at the world. A *science* means a field of study. A *scientist* means anyone doing *research,* no matter how. The new theories and scientific research we hear about on the evening news vary all the way from conjecture and questionable statistical "facts" to newly discovered, experimentally confirmed laws of nature.

Bill Powers writes about different interpretations of theory on an E-mail network:

> Theory, as I see it, purports to be about what we can't experience but can only imagine [with respect to PCT:] (neural signals, functions like input, comparison, output and mathematical properties of closed feedback loops), while evidence is about what we can experience. Both theory and evidence are perceptions, but the way we use these perceptions in relation to each other puts them in different roles.
>
> In the behavioral/social sciences, the word "theory" seems to mean something else: a theory is a proposition to the effect that if we look carefully, we will be able to experience something. A social scientist can say "I have a theory that people over 40 tend to suffer anxiety about their careers more than people under 20 do." The theory itself describes a potentially observable phenomenon. The test is conducted by using measures of anxiety and applying them to populations of the appropriate ages. If we observe that indeed the older population measures higher on the anxiety scale than the younger, we say that the theory is supported—or, as some would put it, the hypothesis can now be granted the status of a theory that is consistent with observation.
>
> This meaning of theory leads to the popular statement that a theory is simply a concise summary of, or generalization from, observations. That definition has been offered by quite a few scientists past and present. I think it misses an essential aspect of science, the creative part that proposes unseen worlds underlying experience. Before the "unseen worlds" definition can make any sense, however, it is necessary to understand, or be willing to admit, that there is more to reality than we can experience. . .

Scientific perspectives

A traditional scientific perspective. It is my impression that most adults take the world for granted. I do. As adults discussing the world, we all have a sense of what some call *objective reality.* We see it in living color, touch it, hear it, smell it, chew it, walk on it, and swim in it. Sometimes we hit it, or it hits us, and it hurts.

Most of us agree that some mental constructs have no equivalent in the physical world we live in. They are what we call *subjective* or *personal.* There is no way to definitively compare one person's subjective impression of things like beauty, marriage, courage, friendship, loyalty, ownership or self-esteem with that of another. What is unclear is where to draw the line between the objective and the subjective.

In electronics, engineers sometimes talk about *black boxes*—electronic assemblies or mechanisms with secret insides but observable and most often very dependable functions. One could say that the function of science is to uncover the secrets of the black boxes we find in nature. In management or behavioral science, the black box is the human being.

An alternative scientific perspective. Instead of taking the world for granted and studying the brain as a black box, we can take the brain for granted and look at the world outside the nervous system as the black box. The challenge now is making sense of that world, starting from the time of emerging awareness in the nervous system of a fetus still in the womb. To see how the nervous system can possibly make sense of its environment, we will need to consider the best available information about neurology, mechanics, physics, chemistry, and biology. We may learn more about the brain looking out from the inside than in from the outside.

Some observations about nerves. Nerves interact with our physiology and the world around us to create the high level human experience we take for granted. Researchers in the fields of biology and neurology tell us that:

1) Nerves are capable of sending streams of pulses through their fibers. Frequencies range from zero to about 1,000 pulses per second. Propagation velocity ranges from 50 to 300 meters per second, which approaches the speed of sound in air.
2) The rate at which pulses are sent appears to be caused by a variety of influences, singly or in combination. Pulses may be originated by

the neuron itself (some continue throughout life), or result from light, vibration, chemicals (hormones), pressure, stretch, temperature, and electricity. Pulses from connecting nerves are another typical source, causing pulses to propagate from nerve cell to nerve cell. A stream of pulses can be called a neural current. Depending on how the neurons are arranged and connected, currents can be added, subtracted, branched, multiplied, integrated, etc., making almost any logical manipulation possible.

We can never know REALITY. Philosophers have argued about what really exists. I accept that the physical world exists, and that we as physical entities are part of and exist in this physical world. The physical world as it exists, I call physical REALITY. I recognize that we are limited in what we can know about the REALITY we are part of. I call the representation we develop in our minds perceptual *reality*.

The complexities of nerves and nerve function are interesting in their own right and will be the subject of detailed research for centuries to come. The intent here is simply to note that all the nervous system can possibly know about its environment (REALITY) are the neural currents travelling in nerve fibers *(reality)*. No organism can possibly have direct knowledge of the world around the brain (REALITY), even though it sure looks that way and many scientists who have not considered this, take for granted that we do. Exhibit 21.

With this realization, it is no longer useful to draw a line between the objective and the subjective. All anyone can know is subjective *reality*. But the dependability—the effectiveness—of a person's personal *reality* varies greatly. Most of us experience it as 100% dependable when dealing with simple physical phenomena. At the same time, we experience it as uncertain when we deal with high-level mental constructs, both in ourselves and in other living beings. Good theory serves to improve the quality of this uncertain *reality* so that we can deal more confidently with the REAL world we live in.

Good theory serves to improve the quality of this uncertain *reality* so that we can deal more confidently with the REAL world we live in.

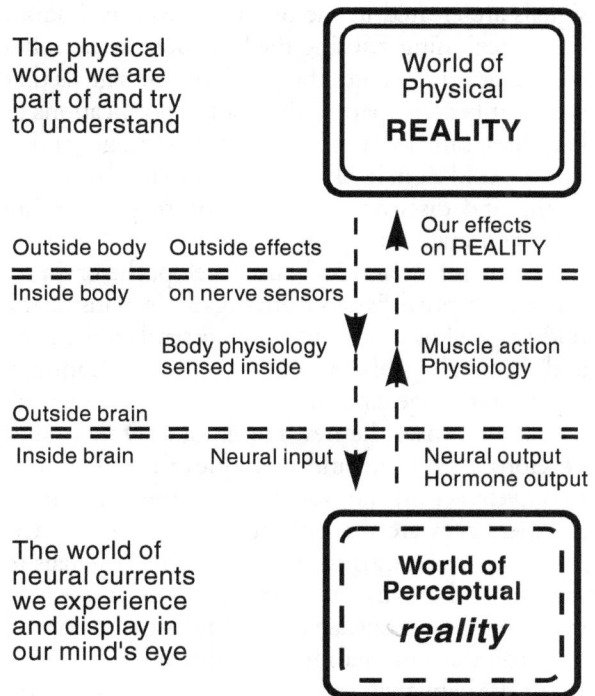

Exhibit 21. REALITY outside. *reality* inside.

Infant perspective: The world as a black box. The challenge for the developing infant is making sense of the currents in its nervous system as signals representing the world outside the brain. The currents originate in a variety of nerve-cell sensors inside the body: in organs and muscles, in eyes and ears, in the nose, mouth and in the skin.

Adult perspective: The brain as a black box. A challenge for life science is to determine the organization of our nerve cells. Taken together, nerve cells make sense of all these currents and develop into a human brain. The adult experiences the world in living color with stereophonic sound—then turns around, takes the world the infant brain has made sense of for granted, *as if* it is experienced directly, and asks questions about the mysterious brain.

Making sense of the black boxes. I certainly don't remember when I became aware of my existence. Adults don't remember much of their early development, but as adults we can observe that the development of infants is rapid. Fetuses still in the womb move about, kick, probably listen and may suck their thumb. A newborn is clumsy at first, but by trial and error finds out what works. When nerves sensing hunger, thirst, heat or cold send signals, other signals are created in the brain, perhaps at random in the beginning, causing the little body to act. If a particular act alleviates the problem, the signals that caused it become part of the brain's specifications to keep itself satisfied; to minimize those hunger, thirst, heat or cold signals. For example, many babies try crying and discover that—as if by magic—crying helps eliminate problems.

As the infant and its brain develop, the brain receives perceptual signals from organs deep inside the body as well as at the surface and sends out neural and chemical signals, causing the muscles to contract, organs to change, and the body to act on the world. The brain senses the new condition. Over time it develops a structure and memories that allow it to effectively act on the world so that the perceptions it experiences are the ones it wants to experience. *The brain acts* (sends neural and chemical signals to muscles and organs) *in order to affect what it experiences*. As time progresses, the baby learns to control its perceptions in ever more sophisticated ways.

As the baby focuses its eyes, coordinates its limbs, enjoys stroking, recognizes sounds and tastes everything it can bring into its mouth, the brain develops a *reality*, an interpretation of the world around the brain. We might say that the baby does scientific research and develops paradigms about the world. In this sense there is no difference between Nobel Prize science and an infant exploring its world. We are all scientists from the beginning of our awareness. But just as Eskimos have many words for different shades of white, we need several words for different shades of theory.

Tools for explanation

Before I discuss theory and explanation, I will review tools we use to describe and explain.

Language: Categorization and generalization. As humans, we benefit from a well developed capability to hear and utter sounds. The infant soon learns to associate sounds with experiences. While some sounds are associated with singular experiences, many words soon come to represent a whole class of experience. The meaning of food, chair, tired, hurt, shoe, walk, sit, and high include several possible configurations of objects, feelings, posture and physical relationships. Language facilitates generalization. Instead of having to duplicate experience, we can describe and categorize experiences.

Logic and Reasoning. Logical reasoning, mathematics and geometry are in a class by themselves. Based on idealized hypothetical postulates, they are logically rigorous. They do not represent physical experience. Therefore, they are not physical sciences, but are valuable as supplements to our descriptive language—tools to manipulate and give precise meaning to descriptions and mechanisms of all kinds, at all levels of sophistication, in all the physical sciences.

Measurement. Measurement is a different kind of tool, linking physical experience with description. Careful measurement has been very important to the development of modern physical science, as exemplified by Galileo's measurements of acceleration.

Statistical Analysis. A special branch of mathematics, statistics is widely used as a diagnostic tool. High correlations between observed variables can prompt guesses about underlying causal mechanisms, which can then be tested to see if the guess is valid. But it is important to recognize the strength as well as the limitation of statistics. In his book *Scientific Explanation and the Causal Structure of the World*, Wesley Salmon (1984) writes:

> Even if a person were perfectly content with an "explanation" of the occurrence of storms in terms of falling barometric readings, we should still say that the behavior of the barometer fails objectively to explain such facts. We must, instead, appeal to meteorological conditions. ... Statistical analyses have important uses, but they fall short of providing genuine scientific understanding .... A rapidly falling barometric reading is a sign of an imminent storm, and it is *highly correlated* with the onset of storms, but it certainly does not *explain* the occurrence of a storm.

Statistical descriptions are useful in terms of populations, whether of people or products, and can be used for prediction in terms of populations. But making decisions about individuals based on statistical prediction amounts to abuse. We call it prejudice. For a discussion of strengths, limitations (why statistical methods are incapable of delivering the secrets of human nature) and misapplication of statistics, please read *Casting Nets and Testing Specimens: Two Grand Methods of Psychology* (Runkel, 1990, 2007).

> The time has come... to put the "cause" back into "because."

Theory, explanation and prediction

Experience. PCT shows that organisms control perceptions, not actions. This explains why organisms do not need to "understand" their environment and why faulty explanations discussed among humans are simply ignored in practice. All an organism needs to do is to pay attention to a perception it wants to control while it acts and remember which way actions influence the variable. An infant lying in the crib reaches for an object hanging overhead. At first the image may be fuzzy and the hand miss the object, but the infant does not give up. It persists and over weeks, months and years learns by trial and error to act on its world so that it can experience it the way it wants to. As adults we have accumulated a large "world" of perceptions which make up and help us function in our individual *perceptual reality*. We call it *experience*. Predators teach their young to hunt through play, demonstration and practice. Consider the tradition in many arts of the master showing the apprentice what and how to perceive: what to look for, how it should feel, sound, smell and taste. We describe only a fraction of our perceptual reality in words. Exhibit 21 and 22.

Predicting from experience. The word hunch captures the idea of theory and prediction in the nonverbal world of *experience* at a very simple level. When we express a hunch we use a few words to summarize a vague or complex notion that we sense, visualize or imagine in the world of perceptual reality, but cannot put into words.

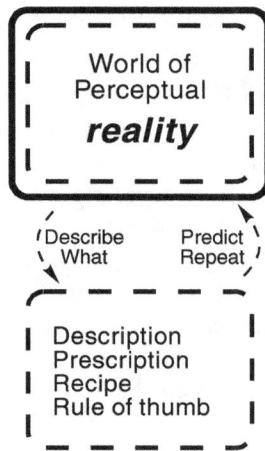

Exhibit 22. Experience and description

Verification and dependability of experience. The words hunch, gut feel, wisdom and mastery suggest degrees of confidence in the predictions we make from experience.

Description. Language allows us to describe our experiences. It becomes possible to learn from experiences of others without having to take the time or suffer the risk of duplicating the experience itself. Our infant becomes a toddler and begins to express experiences in words. Lemons taste sour. Fire burns your skin. Objects fall when you release them. These are simple *descriptions* of phenomena we experience. Exhibit 22.

Prediction from Description. I can now predict that if I bite into another lemon, I will experience sour taste. If I touch fire, I get burned. I predict that when I release an object, it will fall. Predictions are based on regularities; things that usually happen "other things being equal." We use *Rules of thumb*, *Prescriptions* and *Recipes*. Exhibit 22.

Verification and dependability of descriptions. Since descriptions can be shared, they can be compared and the rules can be tried by many people, under different circumstances. We find some rules to be very dependable, while others are uncertain.

Descriptive Non-Explanation. Our preschooler pesters mother with questions. Why, Mother? Why is the Dandelion yellow? Why doesn't the rope break? Because! Because it is strong. Our little scientist is asking questions to make sense of the black box that still holds secrets everywhere you look. Some of mother's answers fit the category of theory I call *Descriptive non-explanation:* The Dandelion is yellow because all Dandelions are yellow. The rope does not break because it is strong, but strong is defined by "does not break." We notice that these are not explanations at all, but restatements or further descriptions of the same experience.

We often explain a phenomenon by using its description, somewhat transformed, as its explanation: You have trouble reading because you are dyslexic. By switching from the English "read" to the Greek "lexia" you make it sound as though you are naming a cause, whereas in fact you are simply repeating the description in a sentence that has the form of an explanation. In one of Molière's plays, a physician explained to a patient that sleeping medication worked because it contained "dormitive principles," where dormir is French for sleep. This term has been used to signify descriptive non-explanation. This is a popular mode of explanation in any field where people keep pestering you for explanations and you find it embarrassing or impolitic to keep saying "I don't know." Exhibit 23.

Exhibit 22 continued:

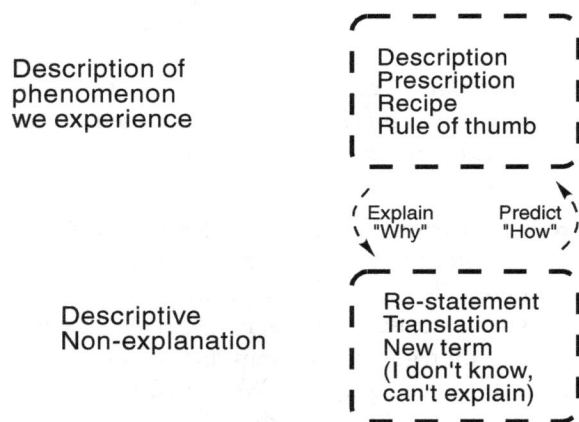

Exhibit 23. Descriptive Non-explanation

There really isn't any difference between descriptions and descriptive non-explanations except for the pretense of explanation and the introduction of a new term. The new term is incorporated in our language.

Causal mechanism. Wesley Salmon (1984) advocates causal mechanisms:

> The time has come, it seems to me, to put the "cause" back into "because." ...The relationships that exist in the world and provide the basis for scientific explanations are causal relations. ...To understand the world and what goes on in it, we must expose its inner workings. To the extent that causal mechanisms operate, they explain how the world works. ...A detailed knowledge of the mechanisms may not be required for successful prediction; it is indispensable to the attainment of genuine scientific understanding.... Explanatory knowledge involves something over and above merely descriptive and/or predictive knowledge, namely, knowledge of underlying mechanisms. ...To untutored common sense, and to many scientists uncorrupted by philosophical training, it is evident that causality plays a central role in scientific explanation. An appropriate answer to

an explanation-seeking why-question normally begins with the word "because," and the causal involvements of the answer are usually not hard to find.

Causal mechanisms suggest the property, structure or functional relationships and interactions of elements below the level of described phenomena. Initially made up in one person's creative imagination, causal mechanisms offer explanations of why and how things happen. The physical sciences, based on causal mechanisms, have progressed far. Exhibit 24 illustrates a series of causal explanations in principle, reaching deep below the surface of the experienced phenomenon and its description.

> A detailed knowledge of the mechanisms may not be required for successful prediction; it is indispensable to the attainment of genuine scientific understanding

Prediction from causal mechanisms. Visualizing the operation of the mechanism in different circumstances, we can predict what effects will emerge. We gain a deeper understanding of what is meant in any given instance when we make a prediction based on some regularity; things that (with high confidence this time) happen "other things being equal." What must be equal? In what way must it be equal? (Ways that allow the mechanism to operate). What does not have to be equal? (Things that do not affect the mechanism). Even a single level of causal mechanism below the level of the phenomenon allows much better prediction.

Verification and dependability of causal mechanisms. We can predict how the mechanism will perform in a multitude of circumstances, even ones we have never experienced before. Experimentation allows us to either reject the proposed mechanism as false and therefore unable to improve our predictions of what will happen, or as 100% dependable. With several levels of such dependable causal mechanisms in the physical sciences, one explaining the other, we have been able to travel to the moon and beyond.

Applications of theory

Causal mechanisms, descriptions and personal non-verbal experience mix when applied. Physical science, rich in causal mechanisms, depends on descriptive empirical data at several levels. A largely descriptive science may have pockets of insight that are of a causal mechanistic nature, whether formalized or not.

To illustrate, I'll share my perspective on applied sciences:

Medicine. Much of medicine is unexplained, and descriptions of symptoms (syndromes) abounds. Much drug research is done by systematic trial and error, just like Edison developed the light bulb. Practicing physicians know that a large part of their job is to comfort and support their patient while nature takes care of healing. Descriptive non-explanations are popular: you have red itchy eyes because of conjunctivitis[1], a red itchy nose because of rhinitis[2], and are cross-eyed because of strabismus[3].

Medicine has made great strides in the last century thanks to the discovery of some causal mechanisms explaining what happens in the body. One example is the discovery of the mechanism of bacterial growth causing the phenomenon of infection. People have learned to avoid harmful bacterial growth through hygiene. Scientists have learned to interfere with bacteria through vaccination and antibiotics, reducing infectious disease. We know that you get other diseases through the mechanisms of virus growth, but have had limited success in interfering with these mechanisms.

When repairing mechanisms of the body, surgeons successfully employ many different causal mechanism explanations derived from the physical sciences.

Mechanical Engineering. Ancient feats of engineering are still admired today: sophisticated compound bows and arrows, ocean crossing canoes, aqueducts, large bridges.

We have few records of exactly how these things were designed and built, but I think it is fair to say that they were based on experience and description, along with some causal mechanism explanations.

---

1 **con·junc·ti′và,** n, the mucus membrane lining the inner surface of the eyelids, covering the front of the eyeball.
2 **rhï°nï′tis,** n. [*rhino-* and *-itis.*] inflammation of the mucous membrane of the nose.
3 **strä°bis′mus,** n. [from Gr. strabismos; *strabizein,* to squint; *strabos,* twisted.] a disorder of the eyes, as cross-eye, in which both eyes cannot be focused on the same point at the same time; squint.

Exhibit 21 and 22 continued:

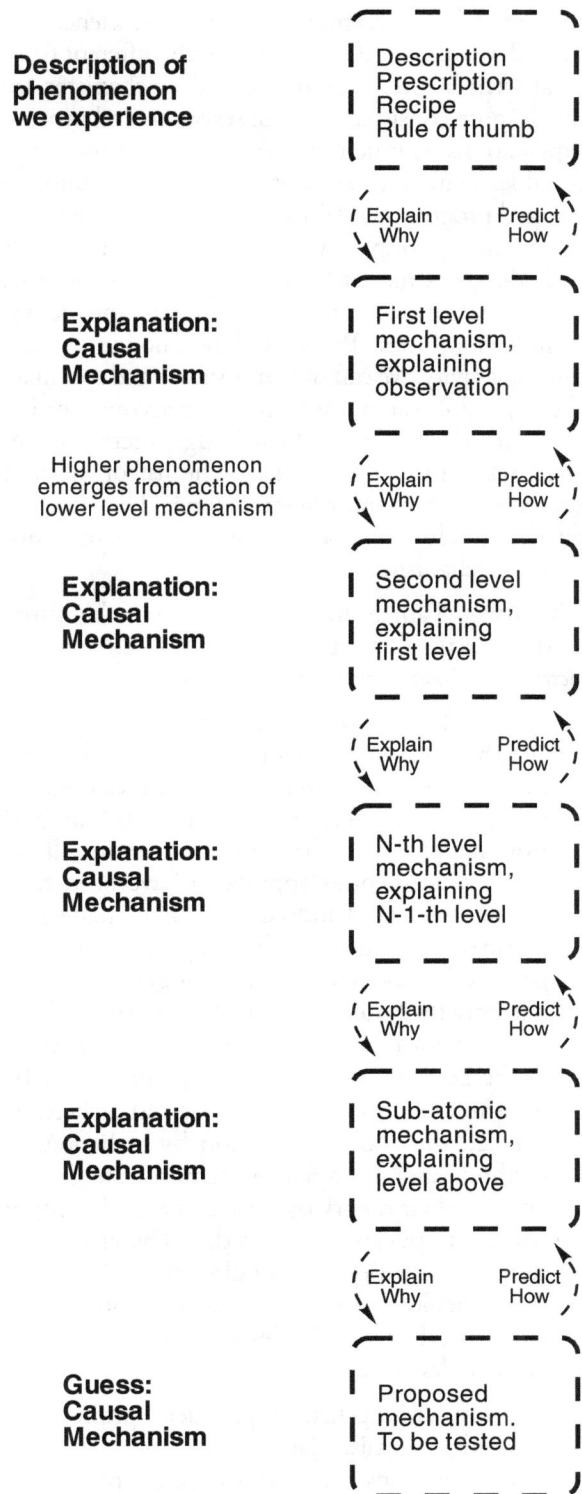

Exhibit 24. Causal mechanisms in depth.

With the advent of the Newtonian laws of nature in the late 1600's and the new rigor of measurement and test of theoretical models, the physical sciences began a development that is qualitatively different from what went before. The new causal mechanisms are much more consistent with observation and provide explanations in much greater depth than had been possible. The last few centuries have seen unprecedented progress in engineering. Causal mechanism explanations, coupled with descriptive data (about such things as material properties) have allowed us to extrapolate from small experiments and design complex machines. But we still have no idea what is causing gravity; we can only measure it and speculate about possible causal mechanisms to account for it.

Railroads, the Golden Gate bridge, aircraft, spacecraft, television, computers... The list of modern engineering accomplishments is long. They depend on the development and verification of in-depth causal mechanisms.

Chemistry. I am aware of three phases of chemistry. To describe the first, let me quote from *Alchemy: Ancient and Modern*, by H. Stanley Redgrove (1911):

> ... we find a school of Arabic alchemy arising in the eighth century A.D. Its inspiration was primarily Hellenistic, and from the contents of many of the texts, much of its theory and practice derived from Egypt. ... The basic idea permeating all the alchemistic theories appears to have been this: All the metals (and, indeed, all forms of matter) are one in origin, and are produced by an evolutionary process. The Soul of them all is one and the same; it is only the Soul that is permanent; the body or outward form, i.e., the mode of manifestation of the Soul, is transitory, and one form may be transmuted into another. ...The old alchemists reached the above conclusion by a theoretical method, and attempted to demonstrate the validity of their theory by means of experiment; in which, it appears, they failed. ...The alchemists cast their theories in a mould entirely fantastic, even ridiculous—they drew unwarrantable analogies—and hence their views cannot be accepted in these days of modern science.

Alchemy in its long history produced products of many kinds—metals, plating, medicine. Alchemy was a descriptive science, a body of prescriptions and recipes based on accumulated experience. The causal mechanism explanations it suggested were failures.

The next phase was dominated by *Phlogiston Theory*. This was an explanation for combustion proposed by Johann Becher (1635-82). It postulated that combustible materials contained an odorless, colorless, weightless (it would rise when released) material called *Phlogiston*. The search for Phlogiston gave direction to much experimentation and by 1775 resulted in the isolation of what was thought to be dephlogisticated air. Today we call it *Oxygen*.

Thus the causal mechanism of Phlogiston failed but was replaced by new explanations for combustion, which we are confident of today. Since the discovery of Oxygen, the science of chemistry has made rapid progress, and is now supported by many additional in-depth mechanisms such as the periodic table of the elements, atomic structure and chemical bonds.

Astronomy. To say that the Sun travels across the heavens in a chariot is indeed to propose a causal mechanism. This and other explanations of celestial phenomena were supplanted by Ptolemy's *Earth centered* model of the universe (c:a AD 140), which placed the Earth at the center of the universe with the heavenly bodies in circular orbits around it. It was apparent that some bodies traveled in reverse periodically, so epicycles, small circular motions, were superimposed on the major circular motion, to describe the apparent paths of individual planets. Over time, this model grew increasingly complex.

Copernicus published an alternate, *Sun centered*, causal mechanism in 1543. This model actually provided predictions which fit observations worse than the existing model. Galileo (1564-1642) developed and published much physical evidence in support of this model. Johannes Kepler (1571-1630) inherited Tycho Brahe's (1546-1601) twenty years of meticulous, descriptive astronomical records, spent additional decades analyzing them, concluding that the planets moved in *ellipses*, not circles. The fit between prediction and data improved. The fit became perfect when Isaac Newton (1642-1727) placed the sun not in the center of the ellipses, as Kepler had done, but in one of two ellipse *focal points*. Newton suggested causal mechanisms to explain how the elliptical motion is created by the heavenly bodies in motion, tugging on each other with (the still unexplained phenomenon of) gravity.

This sequence is interesting as it moves us from an elaborate causal mechanism that appears to work but is fundamentally mistaken, to a fundamentally sound mechanism that appears to work worse, through refinements in several stages to a 100% dependable causal mechanism that today gives us precise results as we continue to map the universe and send spacecraft to the far ends of our solar system.

Psychology: Professional insecurity. Several psychological theories compete for acceptance, with many methods competing for practical use. Many psychologists say that their psychological theories and clinical practice have nothing to do with each other. Scientific psychologists and clinical psychologists have separate societies and professional journals. The diversity of opinion in this field is bewildering. To an electrical, mechanical, or chemical engineer, it would seem strange indeed to be told that there are several electrical, mechanical, or chemical theories, and that practical applications have little or nothing to do with any of the theories.

Psychology: Experience. We all develop an understanding or "feel" for how to deal with people. Most of this "feel" is very personal, intuitive and difficult to express. The style, personality and interpersonal effectiveness that develops from personal experience vary considerably.

Psychology: Description. The vast majority of research in psychology describes apparent phenomena and attempts to relate one description to another by statistical correlation, implying some underlying causal relationship. Such relationships (tendencies, propensities) often are reported despite correlations which sometimes approach pure chance. Over time, stripped of the original uncertainty, many such relationships attain the status of "fact," referred to by subsequent researchers and widely discussed in media. Hidden by statistical summaries are large numbers of counter-examples, where observations are the opposite of reported and popularized "facts." Given more stringent criteria for facts of the physical sciences, where a single counter-example disproves theory, a large number of accepted facts in psychology must be recognized as groundless and simply false. It is unfortunate that psychological descriptive theory is not discarded in the face of counter-examples which disprove it. Instead, uncertain tendencies are used for prediction and judgement of individual behavior. This does not help us resolve conflicts, develop personal relationships, educate capable parents or managers and understand the dynamics of leadership.

Psychology: Descriptive non-explanation. Many popular explanations in this field are descriptive non-explanations. To illustrate, let us take a look at emotion. William T. (Bill) Powers, the creator of PCT, wrote on an E-mail network:

> Emotions are hard to untangle because some people place great value on emotions and don't like to think that emotions might have a rather simple explanation. Emotions, traditionally, are treated as a separate branch of motivation, reaction, or experience, having a somewhat mysterious kind of existence that is neither physical nor mental. Scientists decry arguments that appeal to emotion rather than reason. Their opponents often sneer at emotionless scientists for their coldness or indifference to feelings. Both, when asked to explain what they mean, fall back on descriptive non-explanations.

> Consider the emotion called anger. How do you know when you're feeling anger? In one episode of the television series *Star Trek: The Next Generation*, the android Commander Data asks this question of Geordi, the blind Chief Engineering Officer. In an effort to learn, Data asked Geordi to describe anger without using the word "angry." Geordi (and presumably, the show's writers) are at a loss. "You just—you know—feel *angry*." If you don't know what anger is, how can you understand a description of it? Geordi refuses to fall back on a descriptive non-explanation, and admits that he can't describe anger.

> Well, what does happen when you feel angry? You feel a surge of sensations from your body, and an urge to do something energetic to something. If you have no "self-control" you may well lash out and do damage to something or somebody—anger most often has an object at which you're angry, and it's usually a person.

> The term anger refers to an experience of a surge of bodily feeling and an urge to do something extreme. Anger is just the short way of saying "bodily feeling and an urge to do something." "Anger" isn't an explanation: it's a word referring to a phenomenon that needs an explanation. You don't feel the sensations and the urge to act because of anger, or vice versa. You feel the sensations and the urge to act, or alternatively, you feel anger. The two ways of putting it say the same thing. The word "anger" and the phrase "a surge of bodily feeling and an urge to so something extreme" refer to the same experience. What passes for an explanation is actually a descriptive non-explanation.

Psychology: Failing causal mechanisms. Two major suggested causal mechanism dominate psychology today: behaviorism and cognitive psychology.

Behaviorism[2] suggests that organisms respond to stimuli: What people do depends on what happens to them. Behaviorism includes the ideas of operant conditioning, reinforcement, and affordances; properties of the environment that somehow stimulate us to do what we do. Behaviorism has had a major influence on the psychological understanding of today's teachers and managers. It lays the scientific foundation for our society's love affair with reward and punishment. Data from experiments has varied, so additional, unexplained and unidentified internal and external stimuli have been proposed to account for any mismatch. Critics point out that "behavior" and "stimuli" both are poorly defined.

A major problem with the causal mechanisms suggested by behaviorism is that organisms not only experience stimuli, they create their own. Their behavior obviously, immediately and continuously changes the stimuli that supposedly cause the behavior.

Cognitive psychology describes many phenomena of perception and suggests that behavior is the execution of plans created in our minds.

A major problem with the causal mechanisms suggested by cognitive psychology is that when the brain has to calculate the signals sent to muscle fibers, things will start to go wrong the moment the world around the organism changes. The world may not change in the laboratory, but it sure does in everyday life.

Another problem for contemporary psychological research can only be understood once basic PCT has been understood. The scientific method used in both physical science and psychology simply put is this: Push here and see what happens there. (Change the Independent Variable and observe the Dependent Variable). This method shows what happens naturally with inanimate physical objects, but *not* with animated, active control systems. Control systems resist disturbances! You can learn from the presence or absence of this resistance, but you must understand how a control system works and that you are in fact dealing with a control system. PCT shows that the scientific method has been used incorrectly in psychological research and that all such research must be questioned.

---

2 **bē·hav′iŏr·ĭṣm,** n, in psychology, the theory that all investigation of behavior must be objective or observed as [because] introspection is considered invalid.

Psychology: Present status. Great variation of psychological terminology and interpretation has made it very difficult to agree on consistent descriptions of results. Psychological research is often published despite poor correlations. Studies are rarely replicated to confirm results through independent experimentation, as is routinely done in basic research in the physical sciences. I was startled the first time I was told by a psychologist that psychological theory and practice have nothing to do with each other. Now I understand that this schism is necessary for wise practice based on accumulated experience, since the causal mechanisms offered have not proven valid. But I don't accept that this state of affairs is the nature of science, which the psychologist also claimed.

> Scientists must first understand the new explanation before they can see what is wrong with the old one.

**Psychology of the future:
Successful causal mechanism.**
Organisms live and behave in a world full of influences (disturbances), some of them invisible, (crosswind when you drive), which affect our world (direction of the car) just like our actions (steering) do. These influences should produce instability and failure since they affect outcomes of our actions, but do not. The reason is that our actions automatically compensate for invisible disturbances. The causal mechanisms of psychology discussed above fail because they do not recognize and cannot deal with disturbances in a changing world.

We overcome disturbances and achieve consistent ends by variable means in a changing world because we control. PCT offers a clear and compelling explanation for the phenomenon of control.

HPCT suggests an architecture—an organization in principle of the entire nervous system—suggesting how a system of control systems made up of neurons can develop in the infant and make sense of the world, the black box outside the system.

Neurologists have identified the structure and organization of the neurons surrounding muscle fibers as a control system called the *tendon reflex loop*. A tendon receptor senses tension and sends a perceptual signal (current) representing the tension. A *reference signal*, a signal specifying the momentarily desired tension, arrives through a string of neurons from a higher level in the nervous system. The last neuron in this chain is called the *spinal motor neuron*. The current conveyed through this cell stimulates the muscle fiber to contract, increasing tension at the tendon. A branch of the perceptual signal from the tendon receptor contacts the spinal motor neuron and inhibits its pulse rate. The result of this arrangement is a *comparison* (subtraction) of the stimulating current specifying tension and the inhibiting current reporting perceived tension. This difference is called an *error signal*. In this diagram, the error signal drives muscle contraction directly. In the PCT architecture, a high-level error signal works through other control systems and neural *output functions* to drive *action*. Exhibit 25.

This causal mechanism of neuron interaction explains the lowest level of muscle control we observe when we use muscles in our own bodies and when we experiment on the muscles and nerves of simple animals.

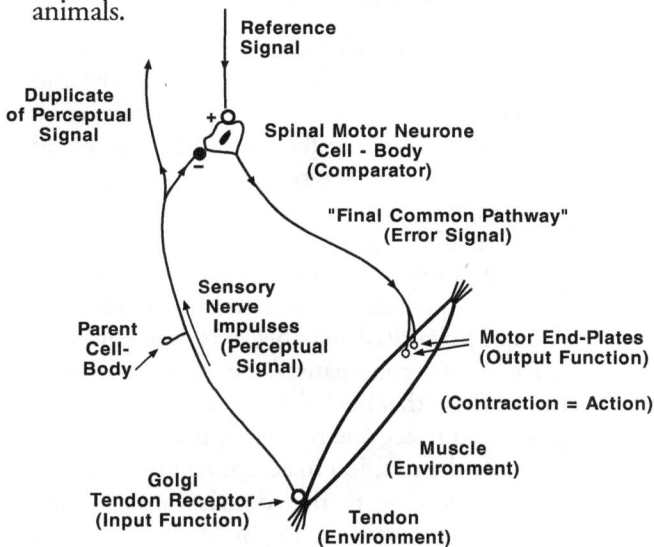

Exhibit 25. The basic first-order control system; the tendon reflex loop. (Powers, 1973).

PCT explains feelings. Bill Powers continues his discussion of emotion:

How would we explain the experience of anger in terms of the PCT control architecture? Clearly, "a surge of bodily feeling" is a perception, and an "urge to do something extreme" implies a control system containing a large error signal. Why, we may ask, would the occurrence of a large error signal in a neural control system be accompanied

by a surge of bodily feeling? One answer that seems reasonable is that the same output of the control system in question that would set reference levels calling for extreme action by the lower motor systems would also set reference levels calling for an altered state of the biochemical systems that support action. Thus we would expect blood sugar to rise, respiration to increase, heart-rate to increase, and so forth—the so-called "general adaptation syndrome." These sudden changes in somatic state can obviously be sensed; they are experienced as bodily feelings.

So when a reference signal is suddenly changed to a relatively extreme value, or a large disturbance suddenly appears, the result is an error-signal-driven urge to change the state of the motor systems and the state of the biochemical systems by a large amount. There is thus a surge of sensation from the body as the biochemical systems are called upon to change to a significantly different state.

Under normal circumstances and in a well-balanced system, the heightened state of preparation of the body is immediately "used up" by the accompanying motor action. There is a momentary sense of elevated somatic state that is simply part of the sensed action. The word "anger" would not be likely to be used to refer to the result.

If, however, the person who experiences the large error has good "self-control," a conflict immediately ensues. One control system receives a reference signal implying an immediate change of state of the whole system, and at the same time a second control system says "No, a civilized person like me does not punch a boor in the nose, whatever the provocation." The "civilized" system cancels the reference signals going to the motor systems, and the punch does not take place.

However, the control system gearing up for the punch is still there, and it is still telling the somatic systems to prepare for violent action. This state of preparedness is now not dissipated by the appropriate motor behavior and disappearance of the error signal; it is maintained by the same error signal that would throw the punch if lower systems were not receiving canceling reference signals from the "civilized" system. The reference signal calling for extreme action is not matched by the appropriate perception, so the urge to act continues and the sensation from the body persists, too. **Now** the person would say "I am angry!"

Moreover, the person would say "I am angry **at him**." The person still wants to see and feel a fist mashing the other's nose, the other person crying out in pain, falling, becoming abject and apologetic and tearful and otherwise suffering all the embellishments of a thoroughly satisfying retribution. All these desires are the immediate source of the reference signal that suddenly changed so as to call for an energetic punch. As long as these desires are in effect, the "civilized" system will have to keep canceling the actual motor reference signals, and the anger and hatred and whatever else we call it will continue. The emotion will persist until the source of the reference signal is turned off. One ceases to be angry when one ceases to want retribution.

This is a PCT explanation of anger that does not rely on a descriptive non-explanation. The same can be done for all the other experiences we label with emotion-names. The feeling component is the perception of a change in the biochemical state of the body, or more generally, somatic state. The goal-component is the reference signal that is calling for both motor action and the somatic state appropriate to the action. If the goal is to get the hell out of there, the same somatic changes take place as in anger, but now the combination of goal and feeling is called alarm, fear, fright, terror, panic, and so on. When the action is prevented from succeeding in achieving the goal, the emotion is felt the most strongly.

Powers concludes:

True connoisseurs of emotion have as large a vocabulary for describing emotions as epicures have for describing tastes and smells. We can speak of feeling annoyed, offended, irritated, provoked, exasperated, angered, incensed, aroused, inflamed, infuriated, and enraged. I've just arranged the terms under "anger" from Roget's Thesaurus in order of increasing error signal and increasing shift in somatic state, as I understand them.

Notice how those adjectives imply the passive voice. It isn't common to attribute emotions to one's own desires. Emotions—particularly the somatic feeling part—seem to arise as though they're being done to us by something else. "You make me angry!" We don't understand where they come from; that's why we need causal mechanisms. In this case, the PCT mechanism tells us we gambled on the wrong voice: we produce our own emotions, which arise from what we want. All these terms should be used in the active voice, which sounds really strange when you do it: "I'm angering at you!"

PCT offers detailed causal mechanisms, subject to refinement in coming decades and centuries. It is possible to generate predictions and effective practices from an in-depth understanding of these causal mechanisms.

Productive and satisfying relationships in the work place, non-manipulative buying and selling in business, loving family relationships, effective education, confident individuals, effective counseling, better understanding of biology, neurology and medicine. The list of improvements will be long. Just like the progress we have already enjoyed in the physical sciences, they will depend heavily on the development and verification of causal mechanisms.

Obstacles to new ideas. Scientific revolutions are not easy. Kuhn (1970) writes:

Because it demands large-scale paradigm destruction and major shifts in the problems and techniques of normal science, the emergence of new theories is generally preceded by a period of pronounced professional insecurity. As one might expect, that insecurity is generated by the persistent failure of the puzzles of normal science to come out as they should. Failure of existing rules is the prelude to a search for new ones. .... Though [scientists] may begin to lose faith and then to consider alternatives, they do not renounce the paradigm that has led them into crisis. ....The decision to reject one paradigm is always simultaneously the decision to accept another, and the judgment leading to that decision involves the comparison of both paradigms with nature and with each other.

The comparison with nature that Kuhn writes about requires the kind of scientific rigor and understanding of causal mechanisms found mostly among those schooled in the physical sciences. Professional insecurity has been present for a long time in the social sciences. A new paradigm is available: The PCT revolution has begun.

PCT: Foundation for physical life science

Exhibit 26 illustrates layers of in-depth explanation in the format of exhibit 24.

At the level of description, PCT deals with familiar phenomena. This can create a problem when communicating about PCT, since some people (not used to causal explanations) look no further and conclude that PCT offers "nothing new."

At the first level of interaction, many lay people have a feel for how individual control (self-direction, freedom) manifests itself in autonomy, conflict and cooperation.

At the second level of explanation, PCT demonstrations of how people can control a single task, acting as an apparent single perceptual control system, are compelling. (Understanding to this level clarifies conflict resolution and personal interactions).

At the third level of explanation, Hierarchical PCT (HPCT) suggests an outline of a hierarchical arrangement of control systems as the organizing principle for the human nervous system. Demonstrations show the operation of such a hierarchy in humans, particularly at lower levels of perception and control. (Understanding to this level clarifies leadership issues).

At the fourth level of explanation, neurologists have identified control systems made up of a few neurons. See exhibit 25.

At the fifth level of explanation, researchers study the structure and interaction of neurons in terms of biology, chemistry and electronics.

PCT and HPCT offer no suggestions for mechanisms behind phenomena such as consciousness, awareness or attention. Understanding the operation of the human mind in greater detail will require research for many years to come, especially at the third through fifth levels of explanation outlined here, including biochemical control systems of several kinds.

It is not necessary to wait for additional research. Even a cursory understanding of the demonstrable concepts of PCT and HPCT offer immediate advantages, as this understanding leads to more effective and satisfying personal interactions.

Exhibit 21 and 22 continued with application to PCT:

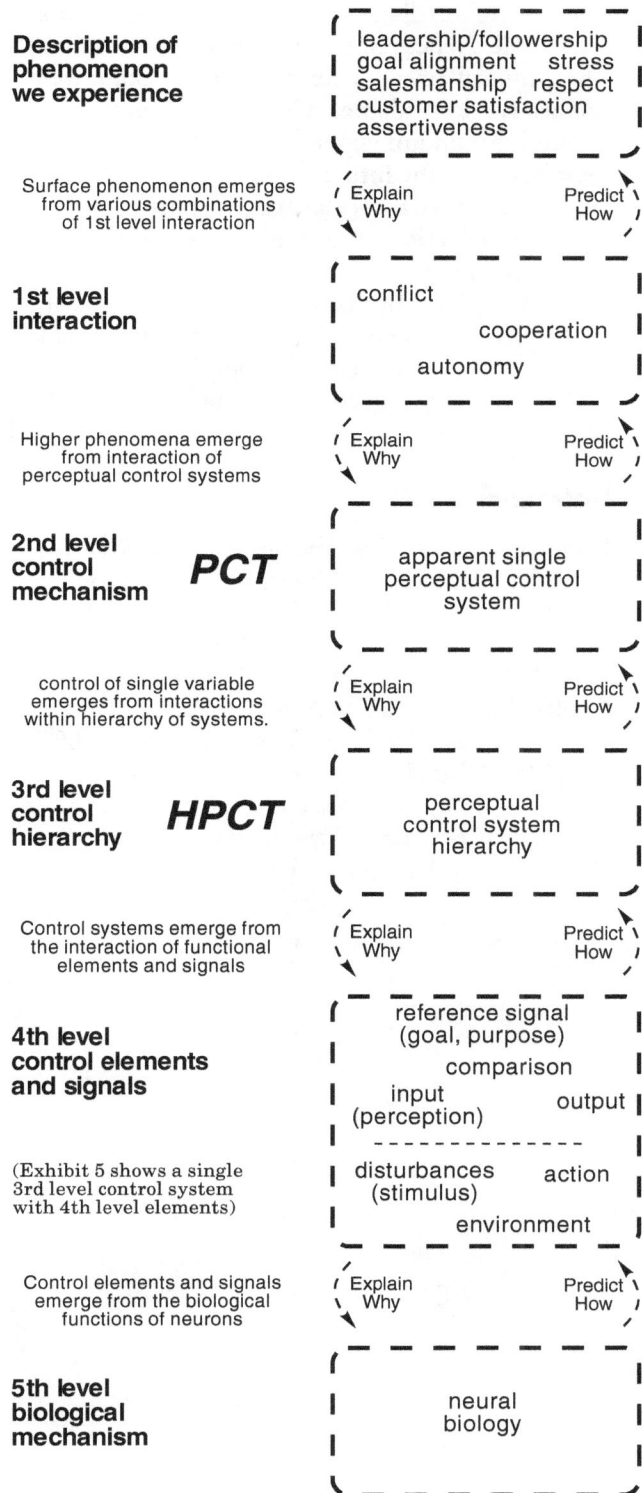

Description of phenomenon we experience
leadership/followership goal alignment stress salesmanship respect customer satisfaction assertiveness

Surface phenomenon emerges from various combinations of 1st level interaction — Explain Why / Predict How

1st level interaction
conflict cooperation autonomy

Higher phenomena emerge from interaction of perceptual control systems — Explain Why / Predict How

2nd level control mechanism — *PCT*
apparent single perceptual control system

control of single variable emerges from interactions within hierarchy of systems. — Explain Why / Predict How

3rd level control hierarchy — *HPCT*
perceptual control system hierarchy

Control systems emerge from the interaction of functional elements and signals — Explain Why / Predict How

4th level control elements and signals
reference signal (goal, purpose) comparison input (perception) output disturbances (stimulus) action environment

(Exhibit 5 shows a single 3rd level control system with 4th level elements)

Control elements and signals emerge from the biological functions of neurons — Explain Why / Predict How

5th level biological mechanism
neural biology

Exhibit 26. PCT psychology: Causal mechanisms in depth.

Conclusion

The point of this discussion of theory and explanation is this: **All sciences of today are *not* created equal. The physical sciences we depend on today were not always dependable. The life sciences we cannot and should not depend on today may become dependable in the future.** The difference lies in the kind and depth of theory and explanation a science is based on. Descriptions in the life sciences are often uncertain to the point of uselessness compared to in-depth explanations based on causal mechanisms in the physical sciences. Progress can best be made when we discover, validate and apply in-depth casual explanations in the life sciences, just like we do in the physical sciences.

References

Kuhn, Thomas S., *The Structure of Scientific Revolutions,* Chicago, IL: University of Chicago Press (1970).

Redgrove, H. Stanley, *Alchemy: Ancient and Modern,* (1911) Reprint by Harper & Row (1973).

Runkel, Philip J., *Casting Nets and Testing Specimens: Two Grand Methods of Psychology* (1990, 2007) Menlo Park: Living Control Systems Publishing.

Salmon, Wesley C., *Scientific Explanation and the Causal Structure of the World,* Princeton, NJ: Princeton University Press (1984).

A COMPARISON —
CHEMISTRY VERSUS PSYCHOLOGY

A comparison — chemistry versus psychology

Perceptual Control Theory (PCT) holds the promise of significant improvements in social, educational, managerial and leadership practices.

This new science is based on engineering principles. It is challenging, because explanations are different from what is understood today. Here is an introduction by way of an analogy to a scientific revolution in another field.

To get a feel for the kind and quality of change the advocates of PCT expect, let us go....

Back to the 16th century:

Imagine that we were born here and now study the science and practice of alchemy (named for the art of making gold and silver). Alchemy is based on practical chemistry know-how, developed by trial an error over many centuries, and incorporates astrology, philosophy and mysticism. As a science it offers descriptions, prescriptions and recipes passed down from past generations of scientists. Alchemy works, and the accomplishments are undeniable. Just look at the great variety of useful products it has given us: metals, metal plating, medicines and much more.

In the 1500's, we live in a society accepting of and dependent on alchemy, where our scientists know what they know, are proud of it, respected, and authorities on their specialty. They write the textbooks used in alchemy school (Gutenberg's printing press is a blessing), referee and edit scientific journals. We cannot imagine a different science with different ground rules, different explanations and much better results, so naturally those of us using alchemy's teachings are proud of what we know and satisfied with the results we get.

Fast forward to the late 20th century:

The science and practice of chemistry is now based on clear engineering principles—what we call causal mechanisms. We have accomplished far more than was possible with descriptions alone. An alchemist transplanted directly from the 1500's, would probably say that there is nothing fundamentally new—she would see that we are still mixing chemicals—until she learned and understood the theoretical difference in the detailed explanations. Scientists can predict results and design new compounds even before they mix chemicals, because they have a carefully tested and validated theory that explains what goes on as the elements interact. When we think of alchemy, we recognize that the scientists who knew what to do in the 1500's, even though they offered what they thought were explanations, had no clear or valid understanding of the underlying processes—how chemicals bond.

We understand now that they could not know in detail why and how their chemistry worked—when it did. Their descriptions have been forgotten and we smile a knowing smile when we hear stories about their quest to turn lead into gold by mixing chemicals. We recognize that it would take more than just a few minutes to explain our causal mechanisms such as atoms and the periodic table of the elements to scientists who were not used to think that way and had never heard of them. No—that is not right—they knew all about atoms, but not in the way we do now. That prior knowledge would only have made it harder for them to hear what we say.

As a by-product of the scientific revolution in chemistry, historians studying the 16th century approaches to metal smelting, alloying etc., can understand why they were successful with some processes but had problems or failed with others.

. . . . . .

Here in the late twentieth century, the sciences and practices of psychology are based on practical know-how, developed by trial and error over centuries. As a science psychology offers descriptions, prescriptions and some theories (where practice often has nothing to do with the theories) passed down from past generations of scientists. Psychology works after a fashion, and the accomplishments seem undeniable. Just look at the dozens of "treatment modalities" used by counselors, scores of leadership programs taught in industry and "common sense" acceptance in our culture.

In the 1990's, we live in a society accepting of and dependent on our behavioral sciences, where our scientists know what they know, are proud of it, respected, and authorities in their specialties. They write textbooks used in schools of psychology, sociology, and management, referee and edit scientific journals. We cannot imagine different behavioral sciences with different ground rules, different basic explanations and much better results, so naturally those of us using the teachings of psychology as we work with people are proud of what we know and satisfied with the results we get.

Fast forward to the 21st century:

The behavioral sciences are now based on clear engineering principles—what we call causal mechanisms, including recognition of and an accurate explanation for the phenomenon of perceptual control. We have accomplished far more than was possible with description alone. A psychologist transplanted directly from the 1990's would probably say that there is nothing fundamentally new—she would hear that we talk about control, perceptions, goals and action—until she learned and understood the difference in the detailed explanations. Leaders, teachers, managers, parents, workers—all are better able to develop satisfying, productive lives, because all have learned Perceptual Control Theory. PCT is so basic to human growth and development, and so easy to understand, that the basics are taught in elementary school. When we think back, we recognize that the scientists who prescribed how to deal with people in the 1990's, even though they offered descriptive non-explanations and what sounded like causal mechanisms, had no clear or valid understanding of the underlying causal relationships.

We understand now that they could not know in detail why and how human relations (courting, parenting, education, supervising, cooperation) worked—when they did. Their descriptive theories have been forgotten and we smile a knowing smile when we hear stories about their quest to shape the behavior of others without regard to their individual wants. We recognize that it took more than just a few minutes to explain causal mechanisms, control of perception and feedback loops to scientists who were not used to think that way and had never heard of them. No—that is not right—they knew all about control and feedback, but not in the way we do now. That prior knowledge only made it harder for them to hear what we say.

As a by-product of the PCT revolution, historians studying the variety of 20th century approaches to education, leadership, and quality management can understand why some seemed successful, why others had problems and what the human costs were.

Back again to the late 20th century:

You have glimpsed the future. Are you satisfied in the present? You can take advantage of PCT without waiting for the whole world to adopt it.

Dag Forssell   April, 1993

BEHAVIOR: THE CONTROL OF PERCEPTION

The basic reference for Perceptual Control Theory
—must reading for serious students of PCT.
Here is a reproduction of the original book cover.

Behavior: The Control of Perception

The book jacket, inside flaps:

BEHAVIOR: THE CONTROL OF PERCEPTION
WILLIAM T. POWERS

"Powers' *Behavior: The Control of Perception* gives social scientists—finally—an alternative to both behaviorism and psychoanalysis. It provides a way, both elegant and sophisticated, to include the basic contributions of both without being partisan or converted. It allows us to bring the soma, culture, society, behavior, and experience into a single framework. We now know much more than we did before this book was published.

– PAUL J. BOHANNAN, Stanley G. Harris Professor of Social Science, Northwestern University; author of *Divorce and After, Social Anthropology,* and other books.

The highly original thesis of this remarkable book is deceptively simple: that our perceptions are the only reality we can know, and that the purpose of all our actions is to control the state of this perceived world. This simple thesis represents a sharp break with most traditional interpretations of human behavior. The theory set forth and developed in detail in this book proposes a testable model of behavior based on feedback relationships between organism and environment, which can reconcile the conflict between behaviorists and humanists and for the first time put us on the road to an understanding of ourselves that is at once scientific and humane.

The model advanced here explains a range of phenomena from the simplest response of a sensory nerve cell to the construction of a code of ethics, using cybernetic concepts to provide a physical explanation not only for physical acts but also for the existence of goals and purposes. A hierarchical structure of neurological control systems is proposed that is at least potentially identifiable and testable, in which each control system specifies the behavior of lower level systems and thus controls its own perceptions.

The model incorporates the "programming" of behavior in the course of human evolutionary history, the nature and significance of memory, and the reorganizations of behavior brought about by education and experience.

Written with verve and wit, with many illuminating examples and interesting thought questions, *Behavior: The Control of Perception* may well prove to be one of the truly seminal works of our time; at least, this is suggested by the distinguished scholars who read the manuscript in advance of publication (see back cover). The book suggests many new interpretations of neurological, behavioral, and social data, an immense range of new experiments that will modify the model advanced here, and much new insight into such crucial psychological and social processes as education, the resolution of conflict, and the problems of mental illness.

ABOUT THE AUTHOR

William T. Powers received his B.S. in physics and did his graduate work in psychology at Northwestern University. He has consulted for The Center for the Teaching Profession, and was formerly Chief Systems Engineer of the Department of Astronomy at Northwestern. He has published articles in psychology, astronomy and electronics, and has invented and designed a number of electronic instruments.

Behavior: The Control of Perception (1973, reissued as paperback in 2005), as well as other books and papers by Powers, is featured at www.livingcontrolsystems.com.

Behavior: The Control of Perception—continued

Back cover:

RUSSELL L. ACKOFF, Silberberg Professor of Systems Sciences, University of Pennsylvania; Past President of the Operations Research Society of America; author of *The Design of Social Research*, co-author of *On Purposeful Systems, fundamentals of Operation Research,* and other books.

"Publication of William Powers' book, *Behavior: The Control of Perception,* is, in my opinion, a major event in the development of the psychology of perception. The completely new approach he has developed using cybernetic concepts cannot help but be seminal, instigating a new and important line of investigation of a wide range of psychological phenomena in addition to perception. His new way of looking at and conceptualizing old things will help to open the way for a series of important discoveries, and these—because of the rigorous framework he provides—are likely to be sounder scientifically than most of the earlier work that they will displace."

DONALD T. CAMPBELL, Professor of Psychology, Northwestern University; Past President Of the Division of Personality and Social Psychology of the American Psychological Association, co-author of *Unobtrusive Measures* and other books and articles.

"Powers' book is, I am convinced, the very best job to date in the application of feed-back theory (servo-system theory, cybernetics) to psychology. Unlike all of its many predecessors, Powers' book comes up with elegant, relevant, and novel detail. It is the first to really capture the promise of cybernetics. It achieves this by bringing to psychology the concept of the 'reference signal' from servo-system theory, and by an explicit hierarchy of 'orders' of control systems."

THOMAS S. KUHN, Professor of the History of Science, Princeton University; author *of The Structure of Scientific Revolutions.*

"Powers' manuscript, *Behavior: The Control of Perception,* is among the most exciting I have read in some time. The problems are of vast importance, and not only to psychologists; the achieved synthesis is thoroughly original; and the presentation is often convincing and almost invariably suggestive. I shall be watching with interest what happens to research in the directions to which Powers points."

JOHN R. PLATT, Research Biophysicist and Associate Director of the Mental Health Research Institute, University of Michigan; author of *Perception and Change: Projections for Survival* and *Step to Man.*

"Powers has made an important new synthesis in applying the concept of hierarchical levels of feed-back-control systems to brain organization and behavior. His ideas throw new light on neural and brain structure, the role of reafferent stimulation in perception and behavior, hierarchical control mechanisms, goal-seeking and feedback at different levels of organization, and epistemology. The book is written in an easy and personal tone with numerous illuminating examples to illustrate the main new points, and with interesting thought-questions at the end of each chapter."

CARL R. ROGERS, Resident Fellow of the Center for Studies of the Person, La Jolla, California; Past President of the American Psychological Association and recipient of its Distinguished Scientific Contribution Award in 1956; author of *Freedom to Learn, On Becoming A Person,* and other books.

"Here is a profound and original book with which every psychologist—indeed every behavioral scientist—should be acquainted. It is delightful to have a person of such varied and unorthodox background come forth with a unique theory of the way in which behavior is controlled in and by the individual, a theory which should spark a great deal of significant research."

PURPOSEFUL LEADERSHIP
SEMINAR INFORMATION

First one-day seminar   - Basic management   85

Second one-day seminar   - Leadership   87

Third one-day seminar   - Technical detail   89

Comments from participants   91

PURPOSEFUL LEADERSHIP™
INSIGHT FOR EFFECTIVE PRACTICE
EDUCATION CONSULTING PUBLISHING

www.purposefulleadership.com

OBJECTIVE

Commitment to common goals, high performance, consistent results and mutual satisfaction.

PURPOSE

A purpose specifies a perception we want. Action makes it so.

LEADERSHIP

"Leadership is the art of getting someone else to do something you want done because he wants to do it."

—*Dwight D. Eisenhower*

Day one:
Basic management

Understand and resolve conflict
Build confidence
Develop productive relationships

Conflict is the root cause of nearly every management problem. It wastes energy and destroys cooperation, teamwork, personal initiative, care, productivity and quality. Failure to resolve conflict results in stress, frustration and resentment, the destruction of personal relationships and turnover of personnel.

In this one-day seminar you learn that we *are* controllers, it is our nature to control, and that our attempts to control others beget conflict. You learn what control is and how it works. You see how control gives rise to conflict or cooperation, depending

on what individuals want and how they see things. Control is not a dirty word. Control is necessary for life and being "in control" or contented is satisfying. It is when others attempt to control us that we resist and dislike it.

You can avoid and resolve conflict by asking questions and offer information so your associate can be more "in control." This builds confidence in your associates and develops caring, productive relationships. The result is mutual satisfaction and committed associates.

What is *Purposeful Leadership*?

Purposeful Leadership is an educational training program which explains, illustrates, demonstrates and applies a functional model of human self-direction. This model expands on what many people already intuitively sense but cannot articulate, because they have never seen it explained.

The model is called Perceptual Control Theory (PCT): a detailed explanation of how thought becomes action, physiology and feelings. Because PCT offers an understanding of the nature, structure and function of the purposeful process all humans unknowingly use to live, it enhances our effectiveness and satisfaction as leaders, managers, salesmen, teachers and friends, both in the workplace and in our personal lives.

Running a company, department or team has been more difficult than it needs to be because we have lacked an understanding of human behavior that actually fits the way human beings work. With PCT, leaders and staff can learn the same proven understanding and effective approach. You deal with your associates at all levels the same way they in turn deal with customers and suppliers. Dealing with people no longer has to be complex and confusing, a matter of luck, a gift, or something best left to specialists.

PCT is not just "another management theory." There are many different management theories, where "theory" means "rules and expectations based on experiences and some suggested explanations."

These kinds of management theories do not always work. They are very different from theories in the physical sciences of today, where "theory" means "in-depth explanation of causal mechanisms, verified in physical experiments." PCT offers testable physical explanations of how behavior results from our personal purposes and perceptions as we interact with our environment. The clear concept of PCT lays a foundation for a new physical science of behavior. This is why applications of PCT can cover much ground, be consistent and effective, all at the same time.

PCT is a science of human self-direction which has been developed, tested, and documented by a multi-disciplinary group of researchers both inside and outside academe.

PCT explains which leadership actions work and **why**. It provides a yardstick by which other management development programs can be measured, because of the "hard" scientific rigor it brings to the "soft" life sciences.

PCT is readily understood and intuitively satisfying. PCT shows that people are purposeful, acting to control their world so they perceive it the way they want to: behavior is the control of perception.

Will Rogers' saying applies to our knowledge of people:

It's not what we don't know that gets us in trouble —it's what we know that ain't so.

PCT greatly simplifies our understanding of motivation and behavior—it helps us see what ain't so, and we can begin to get out of trouble.

What you will be taught

You will be shown a model which gives you clear insight into what makes people do what they do.

With this insight you will see how cooperation and teamwork differ from conflict and coercion. This is not a formula, but a new way of looking at familiar phenomena, which gives a new and deeper meaning to your experience.

Your concerns will be the starting point for a discussion of conflicts. We will show you a new way to think about them, what is required for cooperation, how to prepare, and how to go about resolving them.

You will be taught:

- How to resolve conflict with mutual satisfaction.
- How to develop and sustain personal relationships.
- When helping a person conflicts with respect for the person.
- What control is and how it works.
- How control gives rise to conflict.
- How feelings and stress arise from control and conflict and how to deal with it.
- What is required for cooperation.
- Why people can say one thing and do another. Why it is not necessary to understand why your environment responds to you the way it does—but that you can be more effective when you do understand correctly.
- Why two people can look at the same facts and draw different conclusions.
- The difference between "intrinsic" and "extrinsic" motivation and why nobody can "motivate" another.
- Why "response" to "stimulus" depends on wants.
- Why you cannot tell what people 'do' by watching what they are 'doing.' (Observing behavior alone does not give insight).

This one-day seminar emphasizes the application of *mapping and influencing wants and perceptions:* How to ask questions to guide your associate to consider what he or she wants, what others want, evaluate any conflict, commit to resolve it and help your associate develop an action plan. We introduce a minimum amount of the theory behind it.

You will participate in a demonstration of how a perceptual control system works. Cooperation and conflict are also demonstrated and defined. Perceptions and wants are discussed in detail. We suggest what it means to respect another person, and what it means to be an effective person.

You learn how to *resolve conflicts* within a person as well as between people. You learn how to *build confidence*, and *develop productive, caring relationships*.

We discuss the goals you set if you want to map wants and perceptions with one of your associates. You learn how to be firm, respectful and caring at the same time.

Planning tools are provided for all steps. You plan and practice conflict resolution.

In one day you have learned an explanation for behavior, seen it illustrated, explored the implications and applied it to familiar situations. You leave the seminar with a set of tools so you can apply your new insight to daily challenges.

Agenda:

Introduction
Leadership and Purpose

Understanding
Self-direction & motivation
Perceptions
Wants
Behavior
Conflict
Cooperation
(All above demonstrated)

Values & Qualities
Respect
Effectiveness

Management Application
Problem solving by
Mapping and influencing
wants and perceptions:
Goals
Requirements
Basic methodology

Role play
Conflict resolution:
Manager/employee
or peer/peer.
Build relationships

> Students' interests will determine in which order seminar topics are introduced.

Day two:
Leadership

In this second one-day seminar we build on the first and extend the methodology to performance coaching reviews, teamwork and non-manipulative sales. The theory is explored in more detail, and we discuss how it suggests a structure for mission and vision statements and clarifies TQM.

Participants learn how a leader can guide and co-ordinate efforts of associates towards common goals while allowing them to control their own experiences freely. The result is high performance, consistent results and mutual satisfaction.

What you will be taught

You will see how the basic insight of Perceptual Control Theory provides analytical tools for all interactions between people and suggests uniform, effective, mutually satisfying leadership practice in a wide variety of applications.

Your concerns will be the starting point for discussions of different leadership issues.

The methodology of *mapping and influencing wants and perceptions* is expanded with additional, more detailed consideration of each step.

Performance appraisals. With a slight modification of the mapping methodology, they become a *coaching session.*

Teamwork. We discuss the dynamics of teamwork. You use mapping to obtain commitment to common goals, to resolve conflicts, and frequent feedback to assure high performance with consistent results.

An approach to *non-manipulative selling* follows from the insight and methodology. This means that you can use the same effective approach to deal with your employees at all levels as they in turn use to deal with prospects and customers.

Vision and mission statements. We relate them to the want discussion in the first day and share an example to illustrate the structure we suggest.

> Students' interests will determine in which order seminar topics are introduced.

Performance coaching reviews
Develop effective teamwork
Non-manipulative selling
Vision and mission statements
Total quality management

Total Quality Management programs can be thought of as *social systems.* Social systems are easily confused with control systems, because similar language is used for each and it is tempting to transfer what you have learned about control systems to social systems without thinking carefully about the differences. Nevertheless, an understanding of control systems makes these programs easier to understand, as it maps the relationships between the different elements of a TQM approach.

Agenda:

Introduction
Review

Understanding
Development of human
    understanding and
    how we select wants.
Paradigms and
    scientific progress,
    role of ignorance
Systems thinking
Feelings, Thinking
Leadership:
    Defined, Goals, Planning
Balance

Leadership Applications
Mapping and influencing
wants and perceptions:
    more detailed process
Team building
Performance coaching reviews
Vision and mission statement
Goal structuring
Non-manipulative selling
Social systems:
    Total quality management

Role plays
Performance coaching review
Non-manipulative sales

Day three:
Technical detail

We review and demonstrate technical details of the model. Students gain an in-depth appreciation for the mechanisms behind behavior and can analyze any situation from first principles. Students can diagnose problems never considered by the teacher or discussed in class. The significance of the model is more fully appreciated and participants' confidence in the theory and program is enhanced.

What you will be taught

You find answers to many questions that suggest themselves once you begin to understand how control explains behavior and experience the effectiveness of focusing on wants and perceptions when you work with others. Your concerns and experiences will be the starting point for exploration of the finer points of the model as it relates to leadership issues.

We discuss scientific thinking, modeling and testing of theories; the human building blocks, such as our cellular structure and what is known of our brain structure; the model's suggestion of a hierarchical human control structure; timing of control in a hierarchical structure; how perceptions govern behavior; the model's suggestions for normal operation, automatic control, passive observation, and imagination and thinking. We also discuss the development and change of a living control system: *reorganization.*

You will act out several demonstrations to show that people function in ways that are explained and predicted by the details of this model. We will run several computer simulations and tutorials to help demonstrate and study the properties of a control system, hierarchical control systems and control systems interacting with other control systems in a common environment. You receive a copy of the DOS computer programs so you can study them yourself.

With this considerable detail under our belt, we can discuss and clarify any area of concern.

In the first two seminars we focused on practical applications of the theory. We provided practical approaches to handle conflict and other leadership challenges, while sketching a minimum of theory to support the selection of this particular approach. The advantage of this "show me what to do first" approach to learning is that the teacher can address problems the student is concerned about (the student pays

Purposeful Leadership and Perceptual Control Theory: In-depth understanding

attention) and demonstrate what to do. A possible shortcoming of this approach is that some students may see the program as just another prescription for action, and fail to see the physical validity of the underlying theory.

Because all students will not be interested in more detailed understanding, we have held it for this optional third day. An advantage of the "theory first" approach this third day is that students gain an in-depth appreciation for the physical mechanisms behind behavior and can analyze any situation from first principles. Thus students can think through problems never considered by the teacher or discussed in class. The significance of the model and its range of use is more fully appreciated. A general shortcoming of the "theory first" approach is that few want to study theory for its own sake. But relevance has already been demonstrated by the third day.

Agenda:

Introduction
Review

Understanding
Human building blocks
Brain structure details
Behavior of perception
Examples of hierarchies
   in our minds
Integration of Memory:
   Normal operation
   Automatic mode
   Passive observation
   Imagination
Reorganization:
   Growth and Learning

Demonstrations
Your body and vision
Computer demonstrations:
   Simulations
   Tutorials
   Social interaction

Applications Insight
Want selection
Vision and mission statements
Perceptions

Comments from participants

- This has been an excellent overview of the PCT theory and your hands-on "how-to approach." Your worksheets are excellent and I particularly enjoyed the "enlightenment" of getting into the "student's world." "What do YOU want" is well on its way to a lasting place in my management vocabulary.
- You have done an excellent job of creating a more people centered paradigm for interpersonal interaction.
- The overall concept had great intuitive appeal. It feels right, makes sense, and can be applied in day to day work life and family life. I have seen the personal benefits of this course in my work life.
- Understanding how we all work is useful and practical.
- Love the message and the messengers.
- The "control" mode was excellent. It makes sense.
- The explanation of humans as a control system, their wants are the driving factor behind their actions; that observing behavior alone doesn't give insight, was done excellently.
- Phil Crosby seems to take to the threat out of the manager or supervisor by saying "the problem is not the person but the process"—PCT seems to focus on the person (in a most humane way). You have plugged a big hole for me by showing me how to deal with people problems rather than avoid them.
- Attaining a new methodology that is non intrusive for determining the root cause of a "problem."
- This tool empowers the communication between people, resulting in increased productivity and success for the project and company.
- Your extensive background knowledge and in-depth understanding of PCT shows in everything you do.
- Confrontations with your boss do not have to be one-sided. Solving problems is a lot easier with a plan.
- This course helps emphasize the importance, and shows results of listening and understanding. Creates team work.

- I think PCT has a bright future. It makes sense.
- Shortly after our PCT class I was made leader of our development team. I've been frantically mapping wants ever since.
- The general ideas work for me.
- I've never had heavy involvement in management, supervision or leadership. Your seminar opened that world for me. Many, many thanks.
- It has expanded my awareness of the motivation of others.
- My questions were answered honestly throughout the course. My concern is that others value this program as much as I do.

This seminar was developed in the 1991-94 time frame and presented once on three consecutive Wednesdays to a group of engineers at a high-tech company.

Comments were made by participants on feedback forms filled out at the end of each of the three sessions. I received letters spelling out how the information had been used a year and a half later.

Next time I present a seminar, I want to spend some time discussing with participants what they are most interested in learning from PCT, and how they want to learn it. I expect to present material in a different order with each group of students, in modular fashion, in order to keep the presentation of PCT as interactive as possible.

Dag Forssell

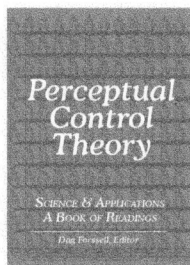

About Perceptual Control Theory (PCT)

It is possible to show quite clearly that organisms behave to control what they experience of the world, an idea that is already striking many scientists as interesting, useful, and quite possibly revolutionary. The books and other resources described here are what people are studying to find out more about this increasingly influential paradigm. Notice what the late Thomas Kuhn said about the first book to introduce what is now called Perceptual Control Theory.

This flyer features books and presentations introducing and applying Perceptual Control Theory (PCT) to various fields.

PCT is taught at the University of Manchester, England. See ***livingcontrolsystems.com*** and ***pctweb.org*** for much more information on the resources featured here, tutorials, student evaluations, academic presentations, videos and more.

For a 20 minute presentation on Perceptual Control Theory, see TEDx Burnley College—Warren Mansell — *Teaching a New Generation about Psychology*: **www.tinyurl.com/TEDx-teaching-psychology.**

Books introducing and applying PCT.   Publishers listed below.[1]

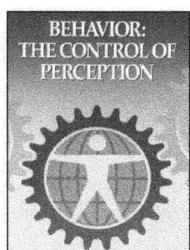

Behavior: The Control of Perception
William T. Powers

978-0-9647121-7-1 (softcover)

Benchmark 1973, 2005

7-5361-2996-3 (softcover, Chinese, 2004)

Powers' manuscript, *Behavior: The Control of Perception*, is among the most exciting I have read in some time. The problems are of vast importance, and not only to psychologists; the achieved synthesis is thoroughly original; and the presentation is often convincing and almost invariably suggestive. I shall be watching with interest what happens to research in the directions to which Powers points.
—Thomas S. Kuhn

Perceptual Control Theory
Science & Applications
—A Book of Readings
Dag Forssell, Editor

978-0-9740155-8-3 (softcover)
978-1-938090-12-7 (hardcover)

LCSP 2008-13

Preview at Google Books. Buy from Internet stores.

This *Book of Readings* provides a sampling of the literature on Perceptual Control Theory, the science and applications to date.

20+ papers cover a broad range of subjects such as feelings, therapy, management, science, and dogma.

Chapters and samples from 16 books on PCT.

Free PDF file at the LCSP website and, as other LCSP books, available for reading at Google books.

Control in the Classroom
An Adventure in Learning and Achievement
Timothy A. Carey

978-1-938090-10-3 (softcover)
978-1-938090-11-0 (hardcover)

LCSP 2012

This new book is a great addition to the educational literature. It introduces educators to the most important and revolutionary new development in psychology in decades, PCT. And it does this in an easy, accessible style. It has something for everyone in education, from pre-school teachers to secondary teachers, as well as their students. Even college instructors and educational policy makers can find much of value in this slim volume. ... Read this book! You'll be glad you did.
— Hugh G. Petrie

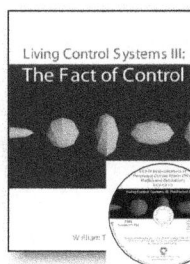

Living Control Systems III
The Fact of Control
William T. Powers

978-0-9647121-8-8 (softcover)

Benchmark 2008   Includes CD

... A unique feature of the book is the accompanying computer programs where Powers 'puts his models where his mouth is,' graphically demonstrating how negative feedback control systems can account for a wide range of goal-oriented behavior. This book is required reading (and computing) for anyone seeking a deep understanding of the behavior of living organisms."
— Gary Cziko

---

1 Benchmark Publications Inc., Kiddy World Promotions B.V., Living Control Systems Publishing (LCSP), Kosmos Uitgevers, MIT Press, New View Publications Inc.

© 2013 Dag Forssell File PCT_literature_resources.pdf from www.livingcontrolsystems.com May 2013

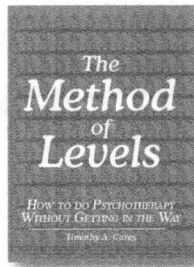

The Method of Levels
How to do Psychotherapy
Without Getting in the Way
Timothy A. Carey

978-0-9740155-4-5 (softcover)
978-1-938090-02-8 (hardcover)

LCSP 2006

Preview at Google Books. Buy from Internet stores.

Tim Carey is the peerless expert on and practitioner of the Method Of Levels (MOL), based on the hierarchical structure of PCT. While working for Scotland's National Health Service he used this approach exclusively with his primary care patients. Some of his colleagues learned MOL from Tim and used it too. MOL achieved a new level of service efficiency as evidenced by the fact that the waiting list went from 15 months when he arrived to less than one month five years later.

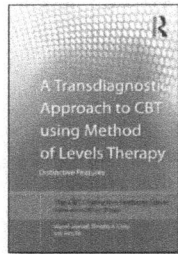

A Transdiagnostic Approach to CBT using Method of Levels Therapy
Warren Mansell, Timothy A. Carey, Sara Tai

978-0-415-50764-6 (softcover)
978-0-415-50763-9 (hardcover)

Routledge Dec 2012

This innovative volume will be essential reading for freshly minted as well as experienced CBT therapists who wish to work using a transdiagnostic approach. Its core principles also apply to counselling, psychotherapy and a range of helping professions. Its accessible explanation of Perceptual Control Theory and its application to real world problems also makes a useful resource for undergraduates, graduates and researchers in psychology.

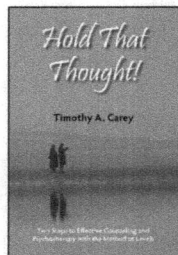

Hold That Thought
Two Steps to Effective Counseling and Psychotherapy With the Method of Levels
Timothy A. Carey

978-0-944337-49-3 (softcover)

New View 2008

Believing that people with psychological problems get themselves better, Carey introduces readers to the Method of Levels, an approach to psychotherapy based on PCT.

Carey's lighthearted style does not obscure his message: that people can change only themselves, and do not need prescriptive solutions from psychotherapists. With lots of examples, Carey shows readers how to find a new perspective on their conflict and ultimately resolve it.

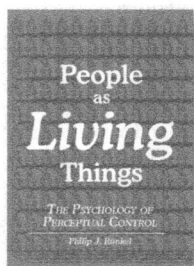

People as Living Things
The Psychology of Perceptual Control
Philip J. Runkel

978-0-9740155-0-7 (softcover)
978-1-938090-01-1 (hardcover)

LCSP 2003

Preview at Google Books. Buy from Internet stores.

Runkel has written a book ... which is at one and the same time: a text book for graduate and undergraduate psychology; an introduction to perceptual control theory (PCT) for the general reader; a paean to William Powers and his achievement—PCT; a memoir about his (Runkel's) exposure to PCT; and an integration of the research and theoretical work on PCT for those familiar with the theory. In my opinion, he succeeds in all these tasks....

—Len Lansky's review: tinyurl.com/lansky-runkel

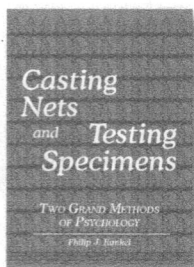

Casting Nets and Testing Specimens
Two Grand Methods of Psychology
Philip J. Runkel

978-0-9740155-7-6 (softcover)
978-1-938090-03-5 (hardcover)

LCSP 1990, 2007

Preview at Google Books. Buy from Internet stores.

A major contribution to the study and practice of socio-psychological research. Runkel's prescriptions understood and followed would revolutionize the behavioral sciences. ... Runkel shows what statistical studies of groups of people, which he calls the method of relative frequencies or "casting nets" can do and what it cannot do: tell anything specific about the nature of individuals. Runkel shows how the scientific study of the individual can get done, what he calls "the method of specimens."

— Bruce I. Kodish

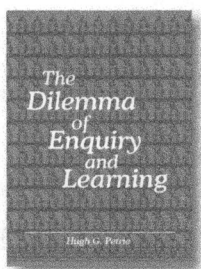

The Dilemma of Enquiry and Learning

Hugh G. Petrie

978-1-938090-06-6 (softcover)
978-1-938090-04-2 (hardcover)

LCSP 1981, 2011

Preview at Google Books. Buy from Internet stores.

I think that this book will be 'compulsory reading' in graduate schools of education around the country, and that it will arouse a vigorous and healthy controversy by shaking people out of unexamined assumptions and compelling them to rethink stale issues in fresh terms.

— Stephen Toulmin

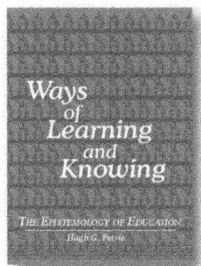

Ways of Learning and Knowing
The Epistemology of Education

Hugh G. Petrie

978-1-938090-06-6 (softcover)
978-1-938090-07-3 (hardcover)

LCSP 2012

For most of his career, Hugh was way ahead of his time. His papers in this volume still are. The role of the evolutionary process of blind variation and selective retention in all knowledge processes and the understanding of behavior as the control of perception are still mostly unknown in mainstream educational research, theory and philosophy. These perspectives, combined with Hugh's analytical skills and accessible writing, lead to some radical (and radically useful) implications for our understanding of the process of knowledge growth and the practice of education.

— Gary Cziko

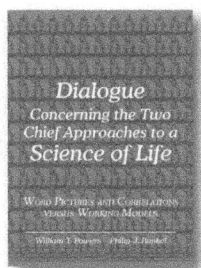

Dialogue Concerning the Two Chief Approaches to a Science of Life
Word Pictures and Correlations versus Working Models

William T. Powers and Philip J. Runkel

978-0-9740155-1-4 (softcover)
978-1-938090-00-4 (hardcover)

LCSP 2011

Preview at Google Books. Buy from Internet stores.

This book holds more than 500 pages of tightly focused, original correspondence between two lucid gentlemen—the creator of PCT, William T. (Bill) Powers, and Philip J. (Phil) Runkel. The significance of the correspondence lies in the subject matter, Perceptual Control Theory (PCT).

The preface and Part II provide
—a brief introduction to PCT (p. 509)
—notes regarding PCT and scientific revolutions
—a guide to resources for your study of PCT

From *Dialogue* / Comments on this volume — the letters and the emerging science

[This volume] provides an outstanding case study of how science develops when real scientists are involved. There are suggestions, descriptions of experiments, computer modeling, explorations of consequences, criticisms, false starts, new breakthroughs, and throughout it all the sense that this is real science in the making. ... It is a must read for anyone who is interested in bringing psychology out of the dark ages and in observing how two outstanding scientists make science really work.

Hugh Petrie, Ph.D. (Philosophy) Professor Emeritus and Dean, Graduate School of Education State University of New York at Buffalo

Bill Powers is one of the clearest and most original thinkers in the history of psychology. For decades he has explored with persistence and ingenuity the profound implications of the simple idea that biological organisms are control systems. His background in engineering allowed him to avoid many of the traps that have victimized even the best psychologists of the past. I believe his contributions will stand the test of time.

Henry Yin, Ph.D. (Cognitive Neuroscience) Professor of Psychology & Neuroscience, Duke University, NC

Bill Powers' work in the 20th century will prove to be as important for the life sciences as Charles Darwin's work in the 19th century. By the time this notion has become common knowledge, historians of science will be very happy with this correspondence between two giants.

Frans X. Plooij, Ph.D. (Behavioral Biology) Director, International Research-institute on Infant Studies, Arnhem, The Netherlands

... When I discovered PCT in the late 1990s, I saw immediately a theory that could bridge the gaps between cognition, behaviour, and motivation by considering them as integral components of a single unit—the negative feedback loop. When I read Powers (1973) further, I realised that these units could be configured in such a way as to model learning, memory, planning and mental imagery. I was 'sold', and since this time I have endeavoured to test and apply PCT within my research and clinical work. It is often difficult for therapists to grasp the notion that there can be a precise, empirical and quantitative model of purposive, humanistic psychology—but here it is.

Warren Mansell, Ph.D. (Clinical Psychology) Senior Lecturer, Chartered Clinical Psychologist, Accredited Cognitive Behavioural Therapist, University of Manchester, UK

© 2013 Dag Forssell File PCT_literature_resources.pdf from www.livingcontrolsystems.com May 2013

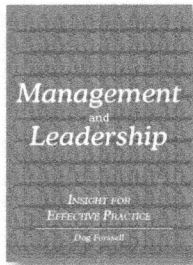

Management and Leadership
Insight for Effective Practice

Dag Forssell

978-0-9740155-5-2 (softcover)
978-1-938090-05-9 (hardcover)

LCSP 2008

Preview at Google Books. Buy from Internet stores.

When i first learned of PCT [back in 1998], I read everything I could get my hands on and your articles, for me, most clearly explained PCT. Somehow, your unique use of language, (perhaps it's more humanizing?) allowed me to understand it better, whereas much that was written (that seems to be changing) is so technical. The result being, if one has not mastered PCT language one becomes lost—at least for a time. Your explanations revealed PCT almost immediately for me.

— David Hubbard, LMHC

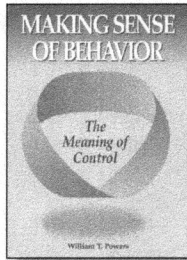

Making Sense of Behavior
The Meaning of Control

William T. Powers

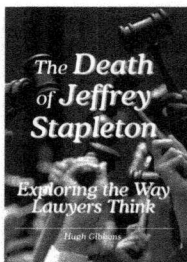

978-0-9647121-5-7 (softcover)

Benchmark 1998-2004

This is the first book on PCT written for "the rest of us." Powers describes in a relaxed, easy-to-read style the fundamentals of this revolutionary theory of the behavior of living organisms— in particular, human beings. This book is for anyone interested in how our systems work and how people interact and why. For researchers new to PCT, a comprehensive reference points to further studies, demonstrations and applications.

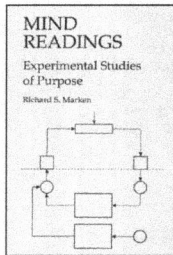

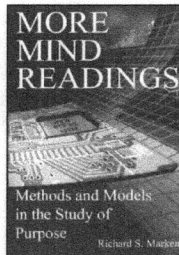

Mind Readings and More Mind Readings
Studies of Purpose

Richard S. Marken

Mind Readings: 978-0-9624154-3-2
More Mind Read: 978-0-944337-43-0

New View 1992 & 2002

[These are books] that can show a willing psychologist how to do a new kind of research. The theme that runs through all these papers is modeling, the ultimate way of finding out what a theory really means. Richard Marken is a skilled modeler, as will be seen. ... He finds the essence of a problem and an elegantly simple way to cast it in the form of a demonstration or an experiment.

— William T. Powers

The Death of Jeffrey Stapleton
Exploring the Ways Lawyers Think

Hugh Gibbons

978-1-938090-08-0 (softcover)
978-1-938090-09-7 (hardcover)

LCSP 1990, 2013

From the Foreword:

Law is the institution that is based upon the assumption that human beings are responsible for their own behavior and the effect of their behavior on others. Perceptual Control Theory, PCT, is the science that explains what behavior is and how it works. The relationship between law and PCT is that simple.

— Hugh Gibbons

A People Primer
The Nature of Living Systems

Shelley A. W. Roy

978-0-944337-47-9 (softcover)

New View 2008

What a blast of a book! Shelley Roy obviously has a deep and clear understanding of Perceptual Control Theory, and her style of presentation shows respect for the intelligence of the reader while at the same time making sure that her message gets across. Shelley successfully suppresses the writer's ego and never condescends—a very nice combination.

— William T. Powers

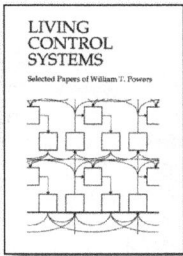

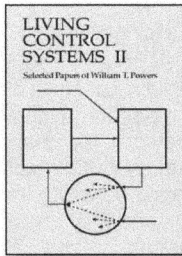

Living Control Systems I and II

Selected papers

William T. Powers

LCS I: 978-0-9647121-3-3
LCS II: 978-0-9647121-4-0

Benchmark 1989 & 1992

Some of the best science is done by people who refuse to take the obvious for granted. Copernicus didn't take the sun's daily trek across the sky for granted, and Einstein didn't take the regular tick of time for granted, and William T. Powers didn't take the appearance of behavior for granted.

— Richard S. Marken

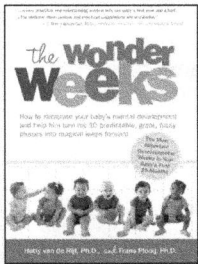

The Wonder Weeks

How to stimulate your baby's mental development and help him turn his 10 predictable, great, fussy phases into magical leaps forward

Hetty van de Rijt, Frans Plooij

978–90–79208–04–3 (softcover)

Kiddy World   2010   In 12 languages

See thewonderweeks.com

The Dutch title for *The Wonder Weeks* can be translated as *Wow, I Am Growing*. Since the original Dutch version was published in 1992, it has sold more than 550,000 copies —in a country of 17 million, one 20th that of the U.S.

This book shows how and when the levels of perception outlined by Hierarchical PCT develop in human infants. The English edition enjoys excellent reviews at Amazon and numerous comments by mommy-bloggers, saying that the predictions about the timing and nature of infant mental development in the first 20 months are right on.

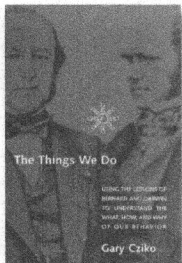

Without Miracles:

Universal Selection Theory and the Second Darwinian Revolution

Gary Cziko

978-0262531474 (softcover)
978-0262032322 (hardcover)

MIT 1995

The Things We Do:

Using the Lessons of Bernard and Darwin to Understand the What, How, and Why of Our Behavior.

Gary Cziko

978-0262032773 (hardcover)

MIT 2000

The inside flap of *The Things We Do* (Complete)

The remarkable achievements that modern science has made in physics, chemistry, biology, medicine, and engineering contrast sharply with our limited knowledge of the human mind and behavior. A major reason for this slow progress, claims Gary Cziko, is that with few exceptions, behavioral and cognitive scientists continue to apply a Newtonian-inspired view of animate behavior as an organisms output determined by environmental input. This one-way cause-effect approach ignores the important findings of two major nineteenth-century biologists, French psychologist Claude Bernard and English naturalist Charles Darwin.

Approaching living organisms as purposeful systems that behave in order to control their perceptions of the external environment provides a new perspective for understanding what, how, and why living beings, including humans, do what they do.

Cziko examines in particular perceptual control theory, which has its roots in Bernard's work on the self-regulating nature of living organisms and in the work of engineers who developed the field of cybernetics during and after World War II. He also shows how our evolutionary past together with Darwinian processes currently occurring within our bodies, such as the evolution of new brain connections, provides insights into the immediate and ultimate causes of behavior.

Writing in an accessible style, Cziko shows how the lessons of Bernard and Darwin, updated with the best of current scientific knowledge, can provide solutions to certain long-standing theoretical and practical problems in behavioral science and enable us to develop new methods and topics for research.

Gary Cziko is Professor and AT&T Technology Fellow in the Department of Educational Psychology at the University of Illinois, Urbana-Champaign. He is the author of *Without Miracles* (MIT Press, 1995).

Oei ik Groei! Voor Managers

Spring door je mentale blokkades

Margreet H. Twijnstra, Frans X. Plooij

978-90-215-5036-7 (softcover)

Kosmos 2011   Dutch only

Dutch title translated:

*Wow, I Am Growing! For Managers
Jump through your mental blocks*

This work features interviews by organizational consultant Margreet H. Twijnstra with 15 top managers who tell of crises in their lives. Frans X. Plooij provides an overview of PCT and explains what is going on during the crisis periods in terms of reorganization, a concept that is integral to Perceptual Control Theory.

© 2013 Dag Forssell File PCT_literature_resources.pdf from www.livingcontrolsystems.com May 2013

Focus on The Method of Levels, MOL

The basic concept of The Method of Levels, MOL, was spelled out in *An Experiment with Levels*, a chapter Powers wrote for the original *Behavior: The Control of Perception*, but which was cut by the book's editors. This chapter is included in *Living Control Systems II* and the concept is again outlined in *Making Sense of Behavior* in the chapter on Inner Conflict. MOL is based on the hierarchical structure of PCT. Dr Timothy Carey was the first to use MOL exclusively in his clinical practice. He refined the approach through research and evaluation, taught it to others, and wrote about his experiences. While working for Scotland's National Health Service from 2002 to 2007, Dr Carey used this approach exclusively with his primary care patients. Some of his colleagues learned MOL from him and used it as well. With MOL, Dr Carey and his colleagues achieved a new level of service efficiency as evidenced by the fact that the waiting list went from 15 months when he arrived to less than one month five years later. During the time he was in Scotland, some clinicians and researchers at the University of Manchester in England became interested in this approach. Dr Warren Mansell (Reader) and Dr Sara Tai (Senior Lecturer) now teach MOL to students, provide training workshops in MOL for other clinicians, conduct research projects investigating different aspects of MOL with postgraduate research students, and use MOL in their clinical practice. Dr Carey continues to use and evaluate MOL as he provides clinical services for the public mental health service in Alice Springs, Australia. Below are some resources you can look up on the web.

Special issue on PCT and MOL
Nine papers on the theory, research and practice of PCT and MOL.
The Cognitive Behavioural Therapist, Volume 2, Issue 3, Sep 2009.
http://journals.cambridge.org/action/displayJournal?jid=CBT

Enabling Flexible Control
Warren Mansell's May Davidson Award Lecture – DCP 2011
... at the British Psychological Society Division of Clinical Psychology Annual Conference in Birmingham on 1st December 2011.
http://www.youtube.com/watch?v=92HaoYRVGcA

What students are saying about PCT
PCT is taught at the University of Manchester as a component of the Psychology degree. See what students have to say here:
http://www.pctweb.org/whatis/students.html

Vignettes on MOL
See YouTube videos at the InsightCBT channel
Look for entries on MOL, Rubber Band and more
http://www.youtube.com/user/InsightCBT

Common Language for Psychotherapy (CLP) procedures
features a definition and explanation of MOL, submitted by Tim Carey and Warren Mansell.
http://www.commonlanguagepsychotherapy.org/fileadmin/user_upload/Accepted_procedures/mol.pdf

Article in Elsevier's *Clinical Psychology Review:*
An integrative mechanistic account of psychological distress, therapeutic change and recovery: The Perceptual Control Theory approach
http://personalpages.manchester.ac.uk/staff/alex.wood/PCT.pdf

Books focusing on or illustrating the Method of Levels include:

The Method of Levels; Hold That Thought; A Transdiagnostic Approach to CBT...; Control in the Classroom. Details on page 1-2

Websites and papers

www.livingcontrolsystems.com

This website features books and introductions to PCT, plus tutorials and simulation programs you can run on Windows.

Recommended downloads at this site::
Powers, William T. (2009).   *PCT in 11 Steps.*
Powers, William T. (2009).   *Reorganization and MOL.*
Soldani, James (1989).   *Effective Personnel Management: An Application of Control Theory.*
Soldani, James (2010).   *How I Applied PCT to Get Results.*

www.pctweb.org

This well developed website is maintained by Dr Warren Mansell from the University of Manchester as an international resource for the dissemination of PCT.

www.mindreadings.com

Rick Marken's website features books, articles and demonstrations you run using your Internet browser.

© 2013 Dag Forssell File PCT_literature_resources.pdf from www.livingcontrolsystems.com May 2013

www.ingramcontent.com/pod-product-compliance
Lightning Source LLC
Chambersburg PA
CBHW080208300326

41934CB00038B/3406